SMART
CARDS

SMART CARDS

A GUIDE TO BUILDING AND MANAGING SMART CARD APPLICATIONS

HENRY DREIFUS

J. THOMAS MONK

WILEY COMPUTER PUBLISHING

John Wiley & Sons, Inc.

New York ◆ Chichester ◆ Weinheim ◆ Brisbane ◆ Singapore ◆ Toronto

Publisher: Robert Ipsen
Editor: Robert M. Elliott
Managing Editor: Erin Singletary
Editorial Assistant: Brian Calandra
Text Design & Composition: North Market Street Graphics

Designations used by companies to distinguish their products are often claimed as trademarks. In all instances where John Wiley & Sons, Inc., is aware of a claim, the product names appear in initial capital or ALL CAPITAL LETTERS. Readers, however, should contact the appropriate companies for more complete information regarding trademarks and registration.

This text is printed on acid-free paper. ⊚

This publication is designed to provide accurate and authoritative information in regard to the subject matter covered. It is sold with the understanding that the publisher is not engaged in professional services. If professional advice or other expert assistance is required, the services of a competent professional person should be sought.

Library of Congress Cataloging-in-Publication Data

Dreifus, Henry, 1960–
    Smart cards : a guide to building and managing smart card
applications / Henry Dreifus, J. Thomas Monk.
        p.   cm.
    "Wiley Computer Publishing."
    Includes index.
    ISBN 0-471-15748-1 (pbk. : alk. paper)
    1. Smart cards.   I. Monk, J. Thomas.   II. Title.
TK7895.S62D74   1997
006—dc21   97-36221
                                                    CIP

Printed in the United States of America

10 9 8 7 6 5 4 3 2 1

For the smart card professionals in the United States who are contributing their time and considerable talent to one of the most exciting technologies since IBM introduced the mainframe computer. The smart card will impact the daily lives of us all, and in ways we are just now beginning to imagine.

CONTENTS

PREFACE

Why This Book Was Written

This book was conceived as a guide to help business managers develop and launch an advanced card system. The concepts discussed provide the framework to develop a feature-rich and sophisticated smart card solution. The concepts and framework are an accumulation of the knowledge that has been gleaned from real systems in place today.

The smart card, a plastic card with an embedded microprocessor silicon chip, can be found in use around the world. Growth has been phenomenal, with over 1 billion smart cards manufactured to date. Soon this card will be used in the United States for credit and debit transactions, for public transportation, and even for buying our morning coffee at the local convenience store.

Even as the financial industry is moving toward adding sophisticated applications like e-cash to credit and debit cards, other sectors are finding benefits in the use of smart cards. Loyalty applications are well known in the retail sector. Many chains issue points each time a purchase is made. Smart cards can provide the electronic equivalent of the S&H green stamps our mothers made us "lick and stick" with dreams of matching sets of dishware.

Health care providers and insurance companies are continually seeking technologies to increase their customer service levels while controlling costs. One application for the smart card is to help reduce the fraudulent claims submitted for payment. In 1995, over $1.2 trillion was paid to suppliers of medical services. Over 9.7 percent of the claims paid were fraudulent. A marginal reduction in fraud would pay for the implementation of a smart card solution many times over. Other applications that would benefit the medical community include tracking immunizations, allergies, drug interactions, emergency medical treatment, and dialysis treatment requirements.

The transportation industry uses the card for ticketless airline payment, subway fare payment, and toll road payments. Each application is designed to improve service and reduce cost for the user and provider alike. It will only be a matter of time before the smart card will be deployed and used in every segment of commerce.

Each smart card solution is unique and involves a myriad of decisions before implementation. Through our consulting practice, we meet executives from firms who are considering smart card applications. In many cases, these executives found out about smart cards from their colleagues in Europe. Although they are captivated by a technology that presents new marketing or manufacturing opportunities for their companies, many of these executives lack an understanding of the big picture. We spend a great deal of time educating new clients through what we call "Smart Card 101." We cover in detail the entire value chain and show how all the participants working together can successfully develop and implement a smart card–based solution. This book is our "Smart Card 101."

Who Should Read This Book

This book will be a valuable tool for chief technology officers and members of their staff, information system architects, project administrators, and system developers seeking to construct or administer smart card solutions in the United States and internationally.

Throughout the book there are suggestions for project managers and team members on how to begin and manage a smart card project, set expectations, and understand the realistic time frames and lead times needed for project budgeting and resource planning.

Business management professionals from the following vertical markets should also be very interested in smart card applications:

- Banking
- Manufacturing
- Entertainment
- Transportation
- Health care
- Government
- Retail

A second group of plastic card, semiconductor, and terminal manufacturers and suppliers will also be interested in this book:

- Visa/MasterCard-issuing banks and equipment suppliers
- Electronic terminal and system access providers
- Smart card manufacturers and issuers
- Telephone operating companies and suppliers
- Internet access providers and e-commerce application developers

How to Read This Book

This book will guide the project manager and decision maker through the entire process of choosing and implementing chip card solutions. The information will be divided into chapters to provide a step-by-step methodology in designing, building, implementing, and managing smart card applications.

Because the book takes an objective, independent perspective in the application development and management phases of a project, we make no specific recommendations for any one supplier, but rather offer a set of experience-driven criteria to assess suppliers within the overall context of smart card applications.

The information in this book has been structured into three parts:

- Introduction
- Design considerations
- Implementation

You may skip over the early information if you are a seasoned smart card veteran, or you may read selectively if you are looking for guidance on specific topics.

Part 1: Introduction (Chapters 1 to 5)

In this section you will find a description of the smart card components, such as types of memory, processors, commands, and modules. These chapters also explain the nuts and bolts of smart card operating systems.

Chapter 1: Evolution of the Smart Card Industry and Market Trends

This chapter outlines the historical foundation for the smart card and discusses market trends, smart card economics, and electronic commerce.

Chapter 2: Smart Card Architectures

This chapter discusses the physical architecture of the single-chip microprocessor card. While all computers contain memory and other commonly understood semiconductor components, their layouts and uses in smart card systems are unique. We introduce smart card hardware, including size and type of memory needed, necessity of a coprocessor for dynamic security authentication, processor clock speed, and proper testing methodologies to ensure that the software works as designed.

Chapter 3: Standards and Specifications

If multiple vendors are to exist and successfully compete in the worldwide market, the size of the card and method of use must be consistent. The International Standards Organization (ISO) has issued a comprehensive set of standards dictating not only how smart cards operate but also the physical parameters (e.g., thickness) for the cards themselves. This chapter discusses the standards for cards and card interfaces in addition to the emergence of industry-specific specifications.

Chapter 4: Security and Cryptography

You will discover how to provide secure, paperless transactions. An entire industry has evolved that is dedicated to money transfers using end-to-end security sessions that are well managed and well controlled by a hierarchical infrastructure.

The smart card will introduce an entirely new approach toward securing information and financial transactions, much of which will be done through dynamic authentication using cryptographic signatures. Sample procedures and architectures contained in this chapter allow readers to determine which, if any, security measures are appropriate for their specific applications.

Chapter 5: Manufacturing the Smart Card

This chapter offers procedures to consider when manufacturing a smart card. These procedures include designing the card face (to accommodate the module) and determining the vendor (or vendors) for chip embedding and card construction. You must also keep lead times in mind to ensure that the manufacturing process remains synchronous with the development of the system. The entire system must be tested and certified on time so that card production can proceed.

Part 2: Design Considerations (Chapters 6 to 9)

If you are already familiar with the nuts and bolts of smart cards, you will want to skip ahead to the section that illustrates the design considerations for implementing a smart card system. Using an example of a successful system as a reference model, we will explain this complex design process and discuss the tools available to facilitate the development effort.

Chapter 6: System Implementation Considerations

Chapter 6 discusses a case study of an electronic purse system that allows customers to buy goods or services electronically. The case study details the trade-offs of various technology choices as well as the different types of smart cards. A section of this chapter is devoted to the tools available to the developer, including off-the-shelf operating systems and component toolboxes and emulators.

Chapter 7: Functional Design and Data Layout Considerations

Chapter 7 continues the case study developed in Chapter 6. A matrix of possible data elements is outlined to help identify the data that is important to specific implementations. We discuss the relative importance of the technical components that must be developed or purchased in order to store and retrieve data, as well as the importance of the design and layout of card memory in sizing the right type of card for the system.

Chapter 8: Operating Systems and the Card-to-Terminal Interface

This chapter discusses the limited number of operating systems available today. Choosing one over another may dictate the chip manufacturer, the security capability, and the card acceptance device (CAD) type for the proposed application or system. Obviously, smart cards will not work by themselves; they must be integrated into an overall system. The card-to-system interface will be through a card acceptance device (CAD) or a reader. Single-vendor versus multivendor CADs are discussed and suggestions offered concerning the functional design of integrating the cards and the CAD terminals.

Chapter 9: System and Data Integrity

As with all information system technologies, smart cards may be either accidentally or deliberately broken or lost; terminals may fail while trying to interface with the cards; even worse, viruses may attack the system. This chapter examines ways to back up card-based information and describes some strategies that can be used to preserve the system's integrity.

Part 3: Implementation (Chapters 10 to 13)

The smart card mosaic is completed by describing the implementation of an application.

Chapter 10: Smart Card Development Skills, Methods, and Tools

The tools and design aids available to develop smart card solutions are discussed, including "mini-application" code fragments that can be put together to script a system. These code fragments have been previously used in commercial implementations and have been tested in millions of transactions. Chapter 10 also discusses the tools and toolkits through an investigation of the pros and cons of the programming languages used in application development.

Chapter 11: Smart Card Testing and Certification

This chapter outlines procedures for testing and certifying the application, chip cards, and terminals that are components of every smart card system. There are established automated and manual procedures used by many smart card providers; this chapter offers tips on how to pretest and bulletproof your system with them. Remember, smart cards are true end-user consumer tools and must work reliably and predictably.

Chapter 12: Implementing and Operating a Smart Card System

The design, testing, and implementation have been completed; it is now time to discuss operating the system. This chapter will lead you through lost cards, monitoring checkpoints in auditing financial applications, upgrading cards in the field to enhance the software or to provide a new set of security keys, and converting from magnetic stripe cards to smart cards.

Chapter 13: The Future of the Industry

This chapter discusses our vision of smart cards and smart card applications, including the ever changing uses of the smart card in today's markets, the addition of new technologies, the ever increasing speed of transactions, and the future of smart cards.

Appendix A: The Smart Manager's Decision Checklists

Even though smart card development projects are very complex, there are a number of decisions that can be made based on the lessons learned during the last several years. Included are decisions regarding applications and systems, manufacturing, system implementation, and functional design. This appendix provides a quick reference for the smart manager to use during the planning process.

Appendix B: Importance of Standards and Specifications—Smart Manager

Chapter 3 discusses the necessity of using ISO standards and industry specifications to provide interoperability between smart card systems. This appendix covers ISO 7816 in greater detail and includes a section of smart card normative references that a smart manager should use during the development cycle.

Appendix C: Encryption Technologies

In Chapter 4, we discuss the use of cryptography to provide a high level of security for smart card transactions. In this appendix, we cover the subject of cryptography in greater detail, with emphasis on RSA, DES, Triple DES, and ECC. For managers with an interest in the Internet, the SET protocol is covered as well.

This appendix also identifies the types of attacks a smart manager should anticipate to ensure that defenses are in place to defeat them.

Appendix D: The Smart Manager's Industry Sourcebook

The smart manager needs to create and maintain an "electronic index" of the key companies that comprise the growing and continually changing smart card industry. This appendix provides a starter set of contacts, including associations, card issuers, card manufacturers, semiconductor manufacturers, system integrators, terminal manufacturers, and a summary of Internet information sites and links.

Appendix E: The Smart Organization

One of the biggest mistakes a smart manager can make is to try to develop and implement a smart card system without making the necessary investments in the organization to make the project successful. This appendix provides the smart manager an organizational structure and job descriptions for the team that must be assembled for a system of 25 to 30 million cards.

ACKNOWLEDGMENTS

This book would not have been possible without the direct involvement of many people. We would like to thank the following:

All of our smart card colleagues who gave so generously of their time and thoughts.

All of our smart card business partners who contributed statistics, charts, figures, and tables. This book is much richer because of your contributions.

Our editors at John Wiley & Sons, Bob Elliott and Brian Calandra, who spent hours helping us collect our thoughts, narrowing our focus, and getting us on some kind of schedule (which in itself was a miracle!).

Our trusted and reliable researchers, Justin Monk, David Belchick, Eric Longo, and Ted Broomfield, who spent many hours on the telephone inquiring, clarifying, sometimes even begging for information, and suffering through numerous edits, rewrites and reorganizations.

And, finally, our "chief editor," Susie Monk and her "chief of staff," Kathy O'Neil-Evans for their creative insight, ability to manage and organize complex tasks, and their insistence on meeting deadlines. They encouraged us at every turn and endured the seemingly endless nights and weekends that it took to write this book. We really appreciate their forbearance.

PART ONE

INTRODUCTION

CHAPTER ONE

EVOLUTION OF THE SMART CARD INDUSTRY AND MARKET TRENDS

Due to vandalism and theft in the early 1980s, France's Public Telephone and Telegraph System began to move to a coinless public telephone system that used "smart" cards to hold a prepurchased value. The smart card—about the size of a credit card—contained a "memory" chip that stored the value. The card could be inserted into a telephone card reader to activate the call and to deduct the cost.

As the use of chip-based telephone cards grew worldwide, a new generation of smart cards began to emerge using an embedded microprocessor to control and safeguard the "exchange" of electronic currency. This new generation of smart card not only serves as a substitute for cash, it also provides added benefits:

- Fraud control for credit and debit cards
- Physical and logical access control for buildings or computer systems
- Storage of emergency medical information
- Unscrambling of cable or satellite "premium" signals
- Ticketless travel on airlines, subways, buses, or trains

To date, over 1 billion smart cards have been issued.

3

The demand for smart cards is growing at a rate of 40 percent per year, as people become more comfortable with the idea of money being stored in a format other than cash and coin. The international credit card associations, which include Visa and MasterCard, have made the ubiquitous credit card a trusted and reliable tool. They accomplished this primarily through developing and implementing an interchangeable format for common transactions.

The implementation of a standard transaction format established the foundation required for merchants to accept both credit and debit cards using the same terminal. To ensure the same level of acceptance for smart cards, the credit card associations formulate technical specifications for cards and card readers to ensure their compatibility. This specification, called the *EMV specification* after the three sponsoring organizations, Europay, MasterCard, and Visa, allows consumers to use smart cards interchangeably. The three franchises provide the technical infrastructure for processing the world's credit and debit cards. The cards themselves are owned by the estimated 22,800 card-issuing organizations around the world.

With demand for smart cards growing, a how-to book on smart cards seemed to be a smart idea. The "why" of smart cards is relatively easy to understand, but the "how-to"—how to plan, develop, implement, and administer smart card systems—is quite a different matter. Performing all these functions effectively is even more challenging!

The purpose of this book is to assist the reader through the process of moving from an idea for a smart card application to a complete, commercial system. If you already have a smart card system up and running, we'll address how you can enhance it. But before we get into the technical discussion, let's lay some groundwork with an overview of smart card market trends.

Market Trends

Worldwide market growth can be attributed to the growth of smart card systems in many different industries, the declining cost of smart cards, and emerging electronic commerce systems.

We'll cover each of these factors in detail in the sections that follow.

Smart Card Growth in Different Industries

The telecommunications industry currently deploys the majority of smart cards for use in GSM (global system for mobile communications) digital cellular, mobile radio, and emerging PCS (personal communication services) systems. These digital cellular technologies use a smart card, or smart card chip, within the telephone handset to enable secure identification and payment of phone calls.

Insurance and health care companies around the world have also adopted smart card technology. In Germany, for example, 78 million chip cards are used to store health care insurance data, including demographic information, payment responsibility, and entitlement benefits. Smart cards are also being deployed worldwide for government-sponsored electronic benefit-transfer and social welfare programs, as well as for electronic identification applications, including driver's licenses, passports, and identification cards that combine smart card and photo identification technologies. Considering that there are hundreds of millions of potential users for each of these applications, a multibillion-dollar smart card industry will shortly be a reality.

Table 1.1 is not an exhaustive list of the major participants in the industry, but it does illustrate the breadth and depth of the players. The leading smart card producers have their roots in Europe. Bull, Gemplus, and Schlumberger (Solaic was acquired by Schlumberger in early 1997) are French corporations, and Giesecke & Devrient and ORGA are German companies. The home countries of smart card reader manufacturers are now truly global, reaching from Japan to the United States.

The greatest industry initiative today is the effort to lower the cost of manufacturing and issuing smart cards.

Smart Card Economics

Telephone cards contain only memory chips with a predefined value stored on them; they aren't able to process data. In many

TABLE 1.1   Advanced Card Producers, Issuers,
and Reader Manufacturers

Producers	Issuers	Readers
Bull CP8 Transac	American Express	Dassault AT
Gemplus DataCard	First Union Bank	DataCard
Giesecke & Devrient	Wells Fargo Bank	Hess
IBM	NationsBank	Hypercom
NBS	Wachovia Bank	International VeriFact
ORGA	Carte Juene/Carta Joven	NBS
Schlumberger-Malco	U.S. Military	OKI
	Shell Oil	Thyron
	Bell Canada	Toshiba
	German government	VeriFone
	Canada Trust	
	Bank of Montreal	
	Toronto Dominion	

countries today, the manufacturing cost for prepaid, disposable telephone cards is less than 50¢ and is expected to fall to around 35¢ before the end of the decade. The microcontroller-based or "smarter" smart cards, on the other hand, cost around $5 to $7 to produce, but are expected to cost between $2 to $3 within the same time period. These smarter cards can be viewed as small "personal computers," with the ability to process data and store information just like the PCs we use in our offices and homes.

Even the more expensive multifunction microcontrollers and cryptographic coprocessors will drop in price. Today, these chips sell for over $10 and provide the capability for multiple applications on one card (such as for debit and credit transactions) and calculate the secret keys necessary for encryption and decryption of data stored on the card.

Several factors contribute to these cost reductions, first and foremost the increase in the number of cards manufactured. Because of improved economies of scale, factories can produce more cards at a lower cost. This phenomenon is the force behind the future of this industry, as demand for cards is expected to double every 12 to 18 months.

Another factor affecting the cost of manufacturing is the semiconductor. All of the chip semiconductor manufacturers have redesigned their products to optimize smart card capabilities on a smaller chip. As the components are reduced in size, the price shrinks at a faster rate than the chip itself! These smaller designs also enable new and better features that enhance the semiconductor's performance.

The third factor is production technology advances that enable faster continuous production of smart cards. The early production lines used a great deal of manual labor, producing chips with questionable reliability due to problems bonding the chip to the card. With newer automated production methods, these problems disappeared and more reliable smart cards were produced at less expense.

Advancements in smart card packaging and chip carrier design also helped make smart cards more cost-effective. In early versions of the cards, chips had a tendency to break or pop out of the card when placed in a wallet or inserted into a card reader. The high failure rates forced manufacturers to design improved micromodules and carriers that would protect the chip.

Today, the chip is wire-bonded to the module to provide the required electrical connections while protecting against stress. Some card producers mill the plastic cards to exact specifications, securing the module in the cavity with a special adhesive. Others use injection molding to manufacture cards and create the cavity in which the module is placed.

Finally, all of the components that form a smart card may now be manufactured under one roof. One company (Siemens, for example) may fabricate the chip, produce the module, and integrate all the pieces in addition to writing the operating system, application, and security software. This vertical integration allows the manufacturer to optimize production with a continuous manufacturing process rather than employing a workstation production environment.

As the cost of the smart card continues to decrease, electronic commerce systems will increase.

Electronic Commerce

Electronic commerce is defined as any monetary transaction that occurs electronically as opposed to the physical exchange of money or checks. In an electronic transaction, e-money is exchanged at the point of interaction after secure information is transmitted across communication lines. Tangible currency is eliminated and accounts are adjusted electronically to reflect the effects of the transaction. For example, an electronic commerce point of interaction may be a pay-per-view or on-demand cable program or a TV connected via modem to the Internet, providing access to a virtual shopping mall. Within this framework, the smart card becomes the electronic key that will unlock a universe of services.

Shoppers in an electronic commerce system do not sign a charge ticket or show a photo ID while shopping; they are viewed as anonymous bits and bytes as opposed to familiar faces at a town store. Within this environment, the smart card takes over the role of consumer identification, electronically secures the link between the merchant and the customer, and authenticates the form of payment.

As the electronic world becomes larger and more impersonal, we will need newer, more efficient technologies to provide reliable identification. Security must improve to lower the risk of fraud without decreasing the performance of the card system.

Implementations of electronic commerce continue to leave the laboratory and enter the pilot-testing stage. Innovative companies like First Virtual Holdings, Cybercash, Wells Fargo, First Union, and (the now defunct) SmartCash have launched electronic commerce projects on university and corporate campuses, in shopping malls, and on the Internet. Consumer banking services have been enhanced through the introduction of branded financial products, such as Visa Cash and MasterCard Cash, to electronically replace the coins and dollar bills used for purchasing small items such as newspapers and sodas. These stored-value card systems, also known as *e-cash* or *electronic purse,* are being implemented around the world and include, for example, NETS in Singapore, NET/1 MegaLink in South Africa, SEMP in Spain, and Proton in Belgium, The Netherlands, Switzerland, Sweden, and the Philippines. (See Table 1.2.)

Before you jump in, it may help to understand the structure of the industry. Figure 1.1 presents an overview of the major participants. The building blocks of the industry are semiconductors, operating systems and masks, chip cards, CADS, and application software. The companies listed are the leaders within their fields. When looking for vendors to help with your chip card project, this figure is an excellent one-stop shop.

The inner circle within Figure 1.1 represents the synergy required between the card issuers, vendors, and project management and systems integration that make smart card systems a reality. The chart illustrates an important concept about smart cards: The smart card itself is only an enabling resource, one component of a very complex system.

Successful Smart Card Applications

The smart card is not science fiction. In the following section we'll show you seven of the most successful implementations of the technology to date, but remember, these only hint at the potential of smart cards. More advanced systems are going on-line as you read this, and the interval between designing a system and releasing a finished product is becoming shorter and shorter, with the reaction time of the market also compressing. Smart card technology, therefore, won't involve designing a system from scratch—there won't be enough time.

Hong Kong's Contactless Transit Card—A Success Story

ERG Ltd., an Australian systems integration firm, created a potential model for future transit applications with its automatic fare-

TABLE 1.2   Examples of Electronic Purses

United States
1996 Summer Olympics, Visa Cash
Manhattan Trial, Chase Manhattan
  Bank
Ohio Dominion College
Oklahoma State University
Florida State University
University of Michigan
Central Michigan University
Western Michigan University
Northern Michigan University
Washington University in St. Louis
Citibank
U.S. Marine Corps, Parris Island
University of Pennsylvania
Movie Gold System
Arksys
Jacksonville Jaguars, Visa Cash
Carolina Panthers, Visa Cash
Lawrence Technological
  University
Louisville Public Schools

Worldwide
Mondex
Visa Cash
Proton

Europe
Eurocheque, Europe
Caf— Project, Europe
Avant, Finland
Quik, Austria

Zeelandkart, Netherlands
SEMP, Spain
SIBS, Portugal
Danmont, Denmark
Zolotaya Korona, Russia

Canada
Restaurants Normadin, Canada

Australia
Freedom, Wizard Card

Asia
Bank of China
Thai Farmers Bank
Nippon Telegraph and Telephone,
  Kochi Prefecture, Japan
Wasida University, Japan
MalaysianCard, Malaysia
NETS, Singapore
CashCard, Singapore

Middle East
Unicard, Isreal

South America
Moeda Electonica Bradesco,
  Brazil

Africa
Mericien Biao
Net/1 Megalink, South Africa

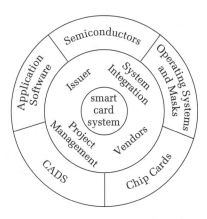

Figure 1.1   Industry overview.

Semiconductor Manufacturers	Chip Card Manufacturers	CAD Manufacturers	Application Software Manufacturers	Operating Systems and MASK Designers
Hitachi	Bull CP8 Transac	Daussalt	ASI (De La Rue)	Bull CP8 Transac
Motorola	Gemplus-DataCard	Hypercom	DigiCash	Gemplus-DataCard
SGS Thompson	Giesecke & Devrient	GPT	FMNT	Giesecke & Devrient
Siemens	ODS	IBM	IBM	ODS
NEC	Orga Card Systems	Ingenico	Siemens Automatix	Orga Card Systems
	Schlumberger/Solaic	OKI		Racom
		Schlumberger		Setec OY
		Toshiba		SCS
		Thyron		Toppan-Moore
		Verifone/H.P.		US3

collection pilot for Hong Kong's various independent public-transport divisions. The system will involve the production and distribution of 3 million reloadable, contactless cards (cards "waved" in front of the reader rather than inserted into it) for use in 400 fare-collection sites located in 14 mass transit stations.

Initially directed by Creative Star Ltd. (a joint venture formed by Hong Kong's Mass Transit Railway Corporation, Kowloon Canton Railway Corporation, Kowloon Motor Bus Company, City Bus, and Yaumati Ferry), the system allows access to all modes of public transit. Its flexibile design will allow stored-value upgrades for use in city parking lots, retail stores, and pay telephones.

First Union's "Spot Card" for the NFL's Jacksonville Jaguars

Among the newest franchises in the NFL, the Jacksonville Jaguars have introduced a disposable, stored-value chip card produced by Schlumberger. The card, with a value of $20, $50, or $100, was implemented in concession and souvenir stands throughout the stadium to speed payment time, reduce errors caused by direct cash transfers, and limit theft. The plan also allows for a general issuance of the card to First Union Bank customers as a reloadable debit card for various retail and transportation use throughout the community.

U.S. Marine Corps, Parris Island

Micro Card Technologies' Morale, Welfare, and Recreation Division Debit Card System has been in operation since 1987 to allow monetary transactions while reducing the use of cash use by Marine recruits training at Parris Island. The card is accepted at several points of sale on base, including Burger King and the bowling alley. The transaction is initiated on-line when the user inputs a PIN number, and the operation is reconciled at a central office. Additionally, remaining funds can be recouped in the event of a lost or stolen card.

Western Governors' Health Passport Project

The Western Governors' Association (WGA) is sponsoring a pilot project to make state health care access and transfer of information for pregnant women, infants, and children easier via smart card technology. It is hoped the card will reduce the stigma of using welfare benefits, empower participants by making them responsible for the information contained within the card, and improve the process of health care by reducing administrative barriers and costs as well as by increasing the timeliness and accuracy of health information. The feasibility study, completed in 1992 by Dreifus Associates and confirmed in 1995 by Price Waterhouse in association with Phoenix Planning and Evaluation, indicated that this multi-state project linking public and private organizations would be technically and economically viable. The WGA HPP program encompasses immunizations, health assessments, diagnosis and treatment, nutritional assessments, food benefits, education and counseling, case management, and prescription review.

1996 Olympic Games

First Union Bank, NationsBank, and Wachovia Bank issued Visa Cash brand smart cards with stored values of $25, $50, and $100 in conjunction with the Olympic Games in Atlanta. By the start of the Olympics, approximately 5000 merchant locations in Atlanta were equipped to accept the cards, which were multifunction cards for pay phones, public transportation, taxis, restaurants, stores, and vending machines.

The State of Ohio

To reduce management costs and fraud, the state of Ohio introduced an *electronic benefits transfer* (EBT) system whereby welfare recipients receive reloadable cards on which the value of food stamps is stored electronically. The consumer spends the value at local grocery stores equipped with readers, thus avoiding the prejudice associated with food stamps, eliminating the administrative process of maintaining paper coupons, and reducing fraud. The state government accepted bids for a privately operated statewide program and has investigated the possibility of creating a multifunction driver's license, state ID, and electronic benefits card.

MARC Project

The U.S. Department of Defense has instituted a multifunction card to replace the various cards used by the military. The Multi-Technology Automated Reader Card (MARC) program was designed to test the feasibility of storing medical and demographic information on a smart card that would replace the dog tag and paper records. The goal of the MARC program is to increase security while decreasing the time and economic impact of data management. The original test, at the U.S. Pacific Command on the island of Oahu in Hawaii, had enough built-in flexibility to allow the addition or modification of functional processes and data. Since its successful demonstration in Hawaii, the MARC card has been adopted for general military use.

CHAPTER TWO

SMART CARD ARCHITECTURES

The smart manager will carefully analyze the overall architecture of the smart card, terminals, and card readers that make up his or her system. The issues facing the manager involve the type and use of memory, communications protocols, and speed, the need for dedicated coprocessors, and the choice of operating system and mask (the part of the application written into ROM). In order to gain maximum functionality within the physically small smart card, trade-offs must be made. These trade-offs occur because the physical limitations of the chip force smart card programs and systems to be written and implemented in slightly different ways than those developed for PCs.

Smart card systems have typically been implemented using assembly-level or another low-level language to program the applications that are loaded inside the cards and terminal readers. These programs allow the card and the reader to communicate with programs written with a higher-level software, such as COBOL or C, on host computers. Improvements in software technology and in the smart cards themselves should encourage the development of more effective ways of creating smart card programs by using interpreters, applets (such as Java), and application development toolkits (to be discussed more fully in Chapter 6, "System Implementation Considerations," and Chapter 10, "Smart Card Development Skills, Methods, and Tools"). Development activities will then be accelerated, resulting in higher reliability and lower cost.

The future for smart cards will mirror the capabilities of today's personal computers. In the future, smart cards will be issued with very little on them other than a core processing engine and an operating system kernel. The cards will be able to accept applications, utility software, and operating system–level software on a dynamic basis through terminals in the field.

Let us look more closely at today's smart card architecture and the layout of single-chip microcomputers. We will begin the discussion with the basic categories of smart cards: memory cards and microcontroller cards, selecting the right smart card option, and the future of smart card systems. (See Figure 2.1 and Table 2.1.)

T I P

Before you begin the design of your smart card system, understand fully the trade-offs between memory types (i.e., ROM, RAM, and EEPROM) for both current and future requirements. (See Table 2.1.)

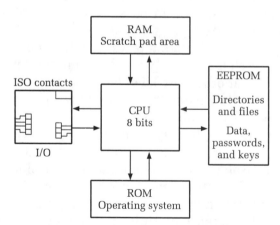

Figure 2.1 Smart card architecture. (Courtesy of Gemplus)

TABLE 2.1 Memory Types

RAM	Random Access Memory
SRAM	Static RAM: contents available during power-on
DRAM	Dynamic RAM: loses content by leakage, has to be refreshed periodically
ROM	Read-Only Memory: contents defined during the semiconductor production process (mask programming)— no alteration possible
WORM	Write Once/Read Many
EPROM	Erasable PROM: erasable under UV light, reprogrammable
EEPROM	Electrically Erasable PROM: erasure of data by electronic signals
Flash	EEPROM: bigger access portions, smaller area

Memory Cards

Memory cards, and all smart cards for that matter, have some form of memory storage. Memory cards are primarily designed for storing information or values and are commonly used for applications such as disposable prepaid telephone cards used in public telephones. These cards are already in use in over 100 countries around the world.

Three major telephone companies in the United States and Bell Canada in Canada are deploying third-generation disposable telephone cards that contain no computing power whatsoever. They are designed only to provide information to a telephone equipped with a special card acceptance device (CAD) or reader. The card provides the telephone an identification number, which may or may not include an embedded PIN number. The PIN ensures that only the telephone operating company that is entitled to take money from the card can read and authenticate the card. The card also stores value that is decremented as telephone calls are made.

Discussion of memory cards can be further divided into general characteristics, EEPROM memory cards, and register memory cards.

General Characteristics of Memory Cards

Memory cards have product options such as register size and memory access time that need to be considered when developing smart card applications. Larger registers allow more data to be accessed or processed at one time. For example, a memory card might use two large registers with faster memory access to meet the required performance criteria. (Of course, the cost of the card is then increased.)

General protocols for communicating with memory cards do not exist. There are no international standards that regulate the communications protocol between the cards and the terminals. In addition, there are no standards for laying out memory on the cards. The memory can be allocated in any form that the hardware and applications designers deem necessary. There is also no uniform definition for the security procedures required to decrement the value of the card or authorize the value remaining on the card. The lack of standards almost guarantees that cards deployed for one system will not work for another.

Memory cards, sometimes referred to as *synchronous cards,* communicate through a defined set of pathways. The synchronous communications process is under the control of the terminal. This method differs from microcontroller-based cards, which are asynchronous in nature; asynchronous cards execute a series of commands and the results are returned through the terminal.

Memory cards also differ from microcontroller cards in their use of electrical contacts and pin assignments. Table 2.2 illustrates the

TABLE 2.2 Memory Map PIN Assignments for Siemens' SLE 4436 Chip

Identification area	64 bit	ROM
Counter area	40 bit	PROM/EEPROM
Anti-tearing flags	4 bit	EEPROM
Authentication Area 1	48 bit	Secret
Data area 1	16 bit	PROM
Data area 2/Authentication Area 2	64 bit	ROM

Courtesy of Siemens

memory layout and pin assignments for a chip manufactured by Siemens AG.

There are two types of memory cards: EEPROM (storage-only) memory cards and memory cards with registers.

EEPROM (Storage-Only) Memory Card

An EEPROM memory card is a storage card with rewritable memory. These cards are used to store information such as a buyer profile for loyalty card programs or database information that might be carried from one application to another. EEPROM memory can be designed as free memory in any format since no standards regulate the allocation of data space inside the card. You can access memory in this kind of card via either a pointer increment process controlling access to specific information or by reading memory in one continuous stream. (See Figure 2.2.)

In the *pointer increment process* a pointer is logically moved 100 spaces to locate the hundredth memory location. The contents of that location, and only that location, are then accessed and read.

When the data is interpreted and updated using the *continuous stream method,* the entire contents of memory are read and then

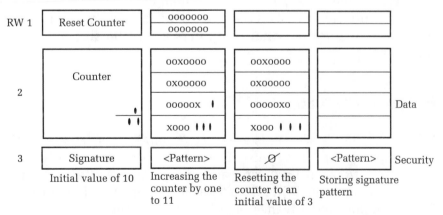

Figure 2.2 EEPROM pointer increment process with security. (Courtesy of SOLAIC)

rewritten to the card. Both techniques are dependent on memory, and they impact not only the size of the chip but the overall application under consideration for the card.

TIP

Maintain a bitmap to account for memory. If this is not done, the cost of the chip may be unnecessarily high as more memory is on the card than is really required for the application.

The second type of memory card is one with registers.

Memory Cards with Registers

These second- and third-generation cards, with very limited memory sizes, use an abacus-style counting method. The abacus method uses a limited-intelligence-register approach. The registers use hard-wired logic to decrement a large number through a series of counter stages with decreasing values. When all the counter stages have been exhausted, there is no money or value left on the card. Since these cards are not rewritable, they are discarded.

Microcontroller cards, on the other hand, are rarely discarded as they are miniature personal computers about the size of a small fingernail.

Microcontroller Cards

Microcontroller-based cards are truly "*smart*" cards. They contain a microprocessor unit, RAM, ROM, mass storage (usually EEP-ROM), input/output hardware, and an operating system or mask. Think of them as the IBM PC that was marketed from 1981 to 1983. Our technology has advanced light-years since that time.

The *microprocessing unit* uses a set of instructions that gives it the ability to process mathematical operations such as addition, subtraction, multiplication, and division. It also has the logic to control the various elements of the device, including in some cases a somewhat simple interrupt control system. An interrupt control system is required when a reader processes information (such as authentication) and needs to signal the card that it is finished and ready to send the results. Interrupts allow the chip to suspend the execution of its program and respond to the device requesting attention.

The chip contains a *random access memory* (RAM) that may vary in size from a low of about 128 bytes to a high of 256 bytes. RAM memory capacity will increase over time as applications require more short-term storage space because of their complexity.

Read-only memory (ROM), which stores the fixed programs, is also included in this fixed allocation of the physical space on the silicon chip. As with RAM, ROM sizes are expected to increase as well.

The last memory area to be included is a *mass storage area,* which ranges from 1 kilobyte to approximately 64 kilobytes today. Mass storage for smart cards is the same as in a PC disk drive, only much more limited, and is used for storing the results of an operation. The information can then be used again in subsequent processing steps. This rewritable memory area will also increase over time as multiapplication systems are implemented and increased amounts of information must be managed by the smart card.

The *input/output hardware* is called the *universal asynchronous receiver/transmitter* (UART). All of the components of the card are connected to the UART, which is in turn connected to the pins that provide physical access to external peripherals such as terminals, printers, and PIN pads. In contrast to the memory cards, which do not have standardized protocols, the layout of the pins for the microcontroller card is standardized according to the International Standards Organization (ISO).

Most cards today require 5 volts to operate. Power for the card is received from the terminal. The next generation of cards will require only 3 volts, thanks to the advancements made in power management for the PC laptop industry.

The last component of a microprocessor is the *mask.* The mask is the tool that actually puts the fixed-program code into the ROM area of the microcontroller card. There are core masks available at reasonable cost from most manufacturers. In fact, most application design tools offer all the basic components for ISO 7816 compliance. Therefore, unless you are going to develop non-ISO-level applications, the specific programming at the base architectural level need not be of great concern.

Spend your time and energy on selecting other options for your smart card systems.

Selecting the Right Smart Card Options

As a system is built, whether it be for health care, financial services, or telecommunications, a specific set of options must be chosen and trade-offs made to maximize the available memory and limited processing abilities of the smart card. Memory cards can be used for specific, relatively simple, prepaid applications like bus tokens or movie tickets. When the application is more complex, a microcontroller-based card system should be considered. The following options should be considered in the decision-making process:

- ◆ Use of dedicated coprocessors
- ◆ Size of the silicon
- ◆ Operating systems
- ◆ File structure
- ◆ Security
- ◆ Terminals
- ◆ Processing speed
- ◆ Testing methodology

Dedicated Coprocessors

Dedicated coprocessors can be incorporated on microcontroller cards to perform more complex functions—for example, fast, dynamic authentication of a security key that nominally should be accomplished in less than 1 to 2 seconds. This hardware is similar in design to floating-point processors and consists of a large parallel set of registers used to perform specific mathematical operations. These registers use one-register arithmetic calculated in single-machine cycles rather than breaking up a number into many little numbers and then running them through an 8-bit microprocessor.

In a microcontroller, a dedicated coprocessor that is designed to process 64- or 128-bit word sizes requires only a limited number of instructions. These instructions are typically arithmetic instructions such as long multiply, long divide, long compare, and various remainder provisions.

There are pros and cons, however, to using a coprocessor. When creating masks for microcontroller applications, developers can rely on ISO standards and protocols to quickly implement the masks and ensure their universal application. Unfortunately, the standards and protocols for coprocessor applications are proprietary and machine-specific. Therefore, the development of coprocessor applications also tends to be more time-consuming than uniprocessor systems. Understanding these trade-offs will help the manager understand the choices both from an economic and an application/capability perspective. Keep in mind, though, that you may be limited by silicon size.

TIP

Remember that the complexity of a system that uses coprocessors is doubled as the second processor also has ROMs and program masks that must be developed and tested.

Silicon Size

There is a practical upper limit to the size of the silicon that can be successfully and reliably embedded in a plastic card. Current technology sets this boundary at about 25 square millimeters, which correlates to an approximate cost of $5 to $7 for a finished card produced in large volume. There are additional factors, including the complexity of the software and the mask hardware, that require trade-offs in the size of the ROM (fixed program) versus the variable-program RAM or EEPROM locations. ROM and RAM are relatively inexpensive when compared to EEPROM. Cost will also be a deciding factor in your operating system choices.

Operating System

Consider, also, the cost of development versus purchase of a smart card core operating system. There are numerous products commercially available (offered by at least 10 software manufacturers) that can run on various subsets of microcontroller chips. (See Table 2.3.) Because this is such a complex decision, operating systems will be discussed in greater detail in Chapter 8.

Once the type of operating system is chosen, file structure should be considered.

File Structure

Once a core operating system has been selected or built, the next design consideration is the file structure. Most smart card operating systems today support only a hierarchical file structure. This file structure is familiar to most programmers and, as in the PC or mainframe environment, carries the same amount of file management overhead when used on a smart card. However, smart card memory is at a premium, so special care must be taken to opti-

TABLE 2.3   Operating System Manufacturers

Bull CP8 Transac	SCS
Giesecke & Devrient	Toppan-Moore
ODS	US3
ORGA Card Systems	Siemens
Racom	Gemplus-Datacard
Setec OY (Finland)	

mize the file design (and many operating systems do not support variable-length records). The same care must be taken during the design of a relational file structure. Research is continuing on object-oriented file structures where data are filed based on their names, not physical location. Space limitations on the card and the cost of an object-enabled terminal are design problems that have not been solved to date.

Security

A major consideration is the method in which the security will be designed into the product. This should include the method for conducting a secure session between a card and a terminal. Security choices must be made concerning whether to require authentication before a transfer of information can occur or to allow some parts of the cards to stay open while others remain closed until they are successfully unlocked using a key-based challenge and response.

Terminals

You'll need to discuss dual use of terminals and the way in which the type of card can be identified when it is inserted into the terminal. Currently, this is not much of an issue, as most of the systems that have been implemented are closed systems with only one type of card. In the future, card-reading terminals like public telephones will need to read both the synchronous memory cards used for pay phones and the asynchronous microcontroller cards issued by Visa, MasterCard, or American Express.

The system design must specify the manner in which a memory card responds to a reset command issued by a terminal. Older memory cards usually respond to indicate that it is a valid chip and that it is a memory card. The response to a reset command varies between manufacturers, and if memory cards are to be used for your specific application, the requirement should be identified in the early stages of the design. Terminal compatibility for ISO-compliant microcontroller-based cards will not be as much of a problem.

T I P

Always plan to accept both memory and micro-controller cards when developing an open smart card system.

Processor Speed

Choices must also be made regarding the microcontroller card processor speed, with the basic trade-off again being one of cost versus application requirements. The ISO standards suggest that processor speed be limited to 3 to 5 megahertz to accommodate the technical constraints of some of the semiconductor manufacturers. Many manufacturers now produce higher-speed controllers in response to market demands, so the ISO standards allow cards to process at higher speeds—but only after going through a handshake process to ensure that both the card and card terminal can support the higher throughput. Five years ago, this was an important factor in the technology, and higher-speed microcontrollers were very expensive. Today, processor speed is not an issue except for some of the small, personal, handheld terminal devices.

Testing Methodology

A final note about card architecture: When a new type of micro-controller design is employed in the application, the methodology to test the software must be developed in parallel. This approach ensures that the software and hardware operate correctly *before* the application is implemented, at which time it becomes very expensive to correct problems.

Trade-Off Decision Matrix

An examination of Table 2.4 illustrates the necessity for an investment in good design. Unlike application development on a PC or

TABLE 2.4   Smart Card Decision Matrix

Feature	Cost	Development Impact
Coprocessor	Higher	Longer time
Chip size/memory	Higher	More complex
Purchased OS	Lower	Limited functions
Developed OS	Higher	Specific functions
File structure	Higher	Flexibility
Security	Higher	More complex
Terminal compatibility	Higher	Flexibility
Communications	Higher	More complex
Testing methodology	Higher	Mandatory

mainframe computer where virtually unlimited resources are available today, working with chip cards is much like data processing in the early 1960s, when machine language and limited hardware resources required detailed design and planning and even more frugal use of the limited physical capability of the hardware. With chip cards, every option must be examined to minimize the cost of the overall system. To assist you in choosing smart card options, we have provided some decision checklists in Appendix A.

Future of Card Architectures

As designs become increasingly complex and silicon manufacturers develop cheaper and more powerful semiconductors, the evolution of smart card technology will accelerate. Today's systems use fixed programs in ROM that cannot be changed once they are installed during the manufacturing process. In the future, it will be possible to load interpretive languages or card-based interpreters into the fixed operating area during the manufacturing process. Applications and applets will then be able to be loaded and unloaded under the control of a terminal as required.

A number of initiatives are under way to develop a *reduced instruction set computer* (RISC) architecture for smart cards. This hardware architecture will take advantage of interpreters and object file management concepts as the machine is fine-tuned to satisfy the requirements of a specific application. Today's assembler-level instructions will become standard long instructions or off-the-shelf instruction sets that will be available to application developers. These architectures will require more memory than is currently available in microcontroller chips.

In choosing a smart card chip or chip set for a given application, you need to keep in mind not only today's applications and needs, but also the growth of that application over time. It is important that an application running on a small 1K microcontroller card can use the same operating system to run on the hardware manufacturer's 8K or 16K product. This strategy allows upgrading to larger memory sizes with full upward compatibility of the basic instructions and instruction sets running on the initial cards.

Upward compatibility of the core application is only one feature required of the emerging new card architectures. Multiapplication capability is quickly becoming a requirement for all developers. As the end users of smart card technology learn to utilize the new service-delivery system, the demand for secure "loyalty" applications (i.e., airline frequent-flyer programs) secure identification and biometric applications, and financial transaction processing will require more integrated circuits within the same physical space.

TIP

An upward migration path must be a part of the design to extend the life of the applications on future chip platforms.

CHAPTER THREE

STANDARDS AND SPECIFICATIONS

What are standards? According to the International Standards Organization (ISO), "Standards are documented agreements containing technical specifications or other precise criteria to be used consistently as rules, guidelines, or definitions of characteristics, to ensure that materials, products, processes and services are fit for their purpose." Standards can be anything from a two-page document to a 1500-page volume. A standard will specify tasks that a piece of equipment must be able to perform or describe in detail an apparatus and its safety features.

Standards govern daily life, specifying the characteristics of many common items: paper size, symbols for automobile controls, currencies and languages, photographic film speed, and even metric screw threads. The format of the credit card, phone card, and smart card that has become commonplace is also derived from an ISO Standard. By adhering to standards for size, optimal thickness, and so on, manufacturers are producing cards that can be used worldwide.

TIP

Standards are not specifications. Even systems that are compliant with standards may not be interoperable.

There is a major difference between a standard and a specification. Typically, a specification is a narrowly defined interpretation of a standard. The formation of a specification usually starts with an existing standard. Certain decisions are made about the various technical alternatives authorized for implementation. The specification defines in detail the technological parameters for the intended application.

When examining standards, you need to understand that they are evolutionary. For example, current standards authorize that the operating voltage for smart cards be 5 volts. However, because most laptop computers and PCs available in the market today commonly operate at 3 volts, the standard has evolved to allow 3-volt operations in smart cards as well.

In this chapter, we discuss plastic card standards, contact card standards, contactless card standards, hybrid card standards, specifications, and compliance with the standards and specifications.

Standards

For the past 15 years, the International Standards Organization (ISO), the electrotechnical counterpart of ISO, the International Electrotechnical Commission (IEC), the European Committee for Standardization (CEN), the European Telecommunications Standards Institute (ETSI), and the British Standards Institute (BSI) have been working aggressively to identify the interoperable or common ways in which cards can be defined for international use. (See Table 3.1.)

Plastic Card Standards

The early application developers adopted the existing standards for cards with magnetic stripes and embossing as their reference points. These standards established the physical characteristics of plastic cards and fixed the location of the magnetic stripe on the back and the embossing on the front of the cards. The chip and

TABLE 3.1 Smart Card Normative Standards

ISO 7810	Physical Characteristics
ISO 7811	Recording Technique Magnetic Stripe and Embossing
ISO 7816	Integrated Circuit Cards with Contacts
ISO 10373	Test Methods
ISO 10536	Contactless Integrated Circuit Cards
ISO 11693	Optical Memory Cards—General Characteristics
ISO 11694	Optical Memory Cards—Linear Recording Method

Source: International Standards Organization

module can be positioned on the card without violating existing standards, thus allowing a smooth migration from magnetic stripe card to chip card.

Smart cards that are embossed (i.e., having a raised surface on the card showing a name or other information) or equipped with a magnetic stripe can be used in any of the three modes. The information on the card can be accessed by reading the chip, swiping the magnetic stripe, or making an imprint from the embossing. Today, three different types of readers are required. However, new readers are now being designed and manufactured that will be able to use any of the three technologies.

ISO 7810 establishes a baseline for the magnetic stripe cards used worldwide for credit and debit applications. This standard defines the location for both the embossing and the magnetic stripe. In addition, ISO 7810 is an important reference standard as it also describes the properties and behaviors of plastic and other materials used to manufacture cards. The standard also defines the physical durability of the cards, including, for example, the amount and duration of its flexibility. Contact cards have their own specifications. (See Table 3.2.)

Contact Card Standards

The second major reference standard for smart cards is ISO 7816, which addresses cards embedded with either microcontroller or

TABLE 3.2   ISO 7810 Physical Characteristics of Cards

Materials: PVC, PVCA, or other materials having equal
  or better performance

Unembossed	*Embossed*
Outer Rectangle:	Outer Rectangle:
Card Width: 85.72 mm	Card Width: 85.90 mm
Card Height: 54.03 mm	Card Height: 54.18 mm
Inner Rectangle	*Inner Rectangle*
Card Width: 86.47 mm	Card Width: 85.47 mm
Card Height: 53.92 mm	Card Height: 53.92 mm

Thickness (all cards): 0, 76 mm +/− 0.08 mm

Source: ISO 7810

memory-only chips. The standard describes the location of the contacts for both types of cards. In the case of the memory cards, however, the standard does not describe the operation of the wires and connections.

ISO 7816 consists of a number of parts (distinct sections). Each part contains specific minimum requirements for the physical characteristics, layout, data access techniques, data storage techniques, numbering systems, and registration procedures. (See Table 3.3.)

TABLE 3.3   ISO 7816 Integrated Circuit Cards with Contacts

Part 1	Physical characteristics (IS)
Part 2	Dimension and location of the contacts (IS)
Part 3	Electronic signals and transmission protocols (IS)
Part 4	Interindustry commands (IS)
Part 5	Numbering system and registration procedure for application identifiers (IS)
Part 6	Data elements for interchange (CD)
Part 7	Interindustry enhanced commands for interchange (WD) (Draft)

Source: ISO 7816

Because ISO 7816 is organized into separate parts, each can be modified and updated as the marketplace and technology advances require. Part 1 describes the physical characteristics (e.g., card thickness) of smart cards and the methods used to test their conformance with requirements. Part 1 also describes the environments in which the cards are expected to operate and the survivability within these environments. This part was recently changed to make it consistent with the published standard on test methods for cards, ISO 10373.

Part 2 addresses the dimensions and the locations of the electrical contacts. Manufacturers are urged to follow the specifications for the location and operation of contacts. By standardizing the physical location, readers can be designed that will work with cards manufactured by all card companies. (See Table 3.4.)

In many cases, card manufacturers have used the contacts to differentiate their products by using unique patterns on the contact surface. The standard specifies only the location for which electrical contact between a card and a terminal need be made. The additional area can be used for larger contact area or for artwork. Part 2 also allows the use of non-ISO contact positions due to terminals that were manufactured and deployed to support smart card systems prior to the adoption of the standard. This variance is particularly important for the early adopters, such as the French banks. In the future, however, these terminals will be replaced with conforming designs. (See Figure 3.1.)

TABLE 3.4 Example of Contact Numbers and Functions (Siemens SLE 4432)

PIN	Card Contact	Symbol	Function
1	C1	VCC	Operation voltage, 5 V
2	C2	RST	Reset
3	C3	CLK	Clock
4	C4	N.C.	Not connected
5	C5	GND	Ground 0 V
6	C6	N.C.	Not connected
7	C7	I/O	Bidirectional data line (open drain)
8	C8	N.C.	Not connected

Courtesy of Siemens

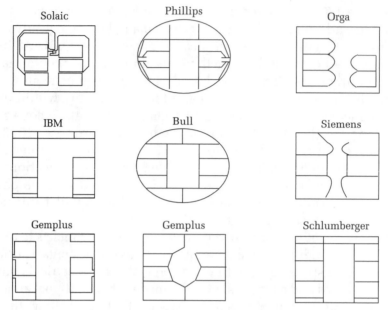

Figure 3.1   Examples of module design.

Part 3 focuses on electronic signals and transmission protocols. This part describes the way in which a card and a terminal communicate. It has recently been updated with new provisions that should be given careful attention. The first of these changes provides the option of using lower operating voltages; as few as 3 volts are now permitted. The second change eliminates the requirement that only contact location 5 be used for programming voltage.

In early smart cards, 21 volts of programming power were required to establish a value in an EPROM cell. With advances in technology, this voltage is no longer required and the use of contact location 5 is now optional. In the future, contact location 5 could be used to implement full-duplex (simultaneous bidirectional) operations.

The reset command brings the card to its initial state. It is important to note that the reset function is the first part of a handshake between a smart card and a terminal. In this handshake, the communication and protocols established depend on the technical capabilities of the card and the card-reading terminal. Part 3 describes the reset command signal and defines the responses that are allowed upon receiving a reset command from a card reader.

The setup bytes transmitted through the reset command are value bytes that describe the capabilities of the card, including the ability to operate at 3 volts or to transfer data using a high-speed, asynchronous, block-transmission protocol. The reset command also identifies the speeds at which the microprocessor can operate. (See Figure 3.2.)

Part 4, first published in September 1995, describes the interindustry commands for the exchange of information between a card and a

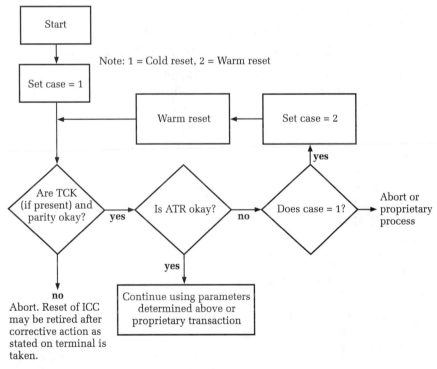

Figure 3.2   Reset command.

card reader. These commands are the basis of a common command set that will allow manufacturers to move away from specific file structures and memory locations containing mandatory attributes or data descriptions. Over time, commands such as READ, WRITE, SET, and RESET will evolve into a set of generic, higher-level commands that will not be application-specific. (See Table 3.5.)

Part 5 discusses the international application numbering system and registration procedure, including the process by which a specific application can be registered. This part provides a framework for uniformity within the industry. Unique applications can reserve identification numbers, just as credit card numbers can be reserved in blocks to identify the issuing bank and the rules to be followed for processing.

TABLE 3.5   Commands and Data Elements

ISO 7816 Commands	ISO 7816 Data Elements
Select_File	Name
Read_Binary	Address
Write_Binary	PAN
Update_Binary	Expiration Date
Erase_Binary	
Read_Records	
Write_Records	
Log_Records	
Update_Records	
Get_Data	
Put_Data	
Verify	
Internal_Authenticate	
External_Authenticate	
Get_Challenge	
Manage_Channel	
Get_Response	
Envelope	

Source: ISO 7816

Currently, the Danish PTT (telephone company) is responsible for the registration process and manages the issued application identification numbers, which are stored on the chip. When the card is inserted in the reader, the identification number discloses what application the card is used for. In today's single-application (closed) systems, this is not an important feature. In the near future, however, the application identification will be important, as many of the cards will be capable of multiple applications and the reader must be able to determine which application is being requested. Many application developers and card manufacturers have registered their specific products or have reserved identification numbers (for example, in health care to provide medical identification or in financial services to perform specific tasks related to stored value, credit, and debit).

Part 6 of ISO Standard 7816 describes the data elements (such as a name, a pin number, an expiration date, and so forth) that can be manipulated by a smart card microcontroller. These data elements become common objects that can be used in the various commands described in Part 4 and can be viewed as a very simple data dictionary.

Part 7, currently in development, is a security architecture and enhanced-command set. Just as the first PC applications using Windows moved on to Windows 95, Part 7 provides upward compatibility for the applications that were implemented under the provisions of Parts 1 through 6. Part 7 describes the additional functions and features that will be available for enhanced commands. Several of these enhanced commands will allow for dynamic execution of downloaded code in an application of the future. The standards for contactless cards are very similar to contact card standards, both in form and content. For a closer look at ISO Standard 7816, see Appendix B, Section 1.

Contactless Card Standards

Contactless card standards are covered in ISO 10536 using Parts similar to those found in ISO 7816. For example, Part 1 describes the physical characteristics of the card. Part 2 describes the dimensions and locations of the coupling areas (location and size of the antenna). Definition of the coupling area is an important topic, given the different types of contactless smart cards that are available in the market. (See Table 3.6.)

The first type of contactless card is very close proximity (less than 1 millimeter) cards, in which the reader/writer coupling devices in the card and the terminal must be very precisely aligned. (Many of these cards allow less than 2° of variance from a vertical alignment with the terminal.) The second contactless card type is a close-proximity card where reader/writer short-interval coupling is also alignment-sensitive. Close-proximity cards require a specific orientation of the card to the reader at a very close distance of 1 to 2 millimeters up to a centimeter.

The third type of contactless card allows remote coupling. Remote-coupling cards can operate within distances of a few centimeters to as many as 3 to 5 meters. The orientation of the card is not important—top up or down—except that it must be perpendicular to the read/write field.

Remote-coupling cards are further described in ISO 1443 which also parallels ISO 10536 for Parts 1 and 2 (dimensions and loca-

TABLE 3.6   Three Types of Contactless Cards

Immediate proximity
- Less than 1 mm distance from reader
- Less than 2° variance from vertical

Close proximity
- Between 1 and 2 mm distance from reader
- Specific orientation

Remote coupling
- Between 3 and 5 mm distance from reader
- No required orientation

Source: DAL

> **TIP**
>
> Of the contactless smart cards described in ISO 10536 we believe that remote-coupling cards will become the most popular commercially and will be used in many ticketless travel applications such as toll roads, airline "electronic tickets," and public transport.

tions of coupling areas). (See Table 3.7.) Part 3 describes the electrical signals and reset procedures that are similar to the contact card reset process. However, contactless cards may not have the one-to-one relationship with a reader that a contact card has. Therefore, to perform the handshake when multiple cards are present in the radio frequency (RF) field, contactless cards have to time themselves. Contactless card manufacturers have agreed on 9600 baud during the first handshake, at which point answers to reset bytes are passed between the card and the card reader. After the reader and card understand each other's capabilities, the session will continue at a baud rate that both can handle.

When several cards are present in the RF field, anticollision methodologies are also employed, similar to those used for local

TABLE 3.7   ISO Contactless and Remote-Coupling Standards

10536 Contactless IC Cards
Part 1: Physical characteristics
Part 2: Dimensions and locations of coupling elements
Part 3: Electrical signals and reset procedures

14443 Remote Coupling Communication Cards
Part 1: Physical characteristics
Part 2: Radio frequency interface
Part 3: Transmission protocols
Part 4: Transmission security features

Source: ISO 10536, ISO 14443

area network (LAN) topologies to assign read/write priorities based on the serial number of the card.

Part 4 discusses the transmission protocols and describes the answer to the reset handshake in greater depth. While Part 3 is oriented toward the way in which a reset signal and condition is handled in a card, Part 4 deals with the issues required to prevent collisions between transmissions when more than one card is present in the read/write field of a terminal. The transmission protocols employed by contactless cards are very similar to the protocols defined for contact cards. Unfortunately, there are no standards developed for hybrid cards.

Hybrid Card Standards

Hybrid (sometimes called "combi") cards contain both a contact and an RF means to communicate. The card is read either by inserting it into a reader or by waving it through a radio frequency field. Most of the systems that use this hybrid technology usually revert to the card contacts for connection and will use the conventions for this means over those established for contactless communications.

Since most of the contactless cards do not contain batteries or ways to store energy, they must receive their operating energy from the transmitting radio frequency field in which the card is operating. The transceiving antenna in a contactless card employs loop-coil technology to power the card. It is for this reason that most of the systems using radio frequency cards tend to be simpler in nature and operate at slower data rates between the card and card reader.

The source of power is not the only reason that contactless cards operate at slower speeds. Distance becomes a factor when the card has to transmit information over a distance of a few centimeters up to 0.3 meters or more. A second factor is the amount of energy available from the RF field. The current remotely powered loop-coil technology is capable of producing only limited power, so the card is restricted to slower microprocessor speeds. The slower CPU cycle time restricts communications, and the card cannot exchange data as fast as contact cards, which can take power directly through their contact interface.

Typically, in developing applications, you will not be working at these lower-level command sets unless you are developing your own protocols and operating systems. Therefore, it is important to have a good understanding of the manufacturer's level and degree of compliance with the ISO standards.

TIP

All suppliers should be qualified for their level of compliance with standards, especially ISO 7816.

Always keep in mind that standards are not rigid rules. You don't have a tight specification that mandates the manner in which a card will work. For example, an ISO standard card can operate at either 5 or 3 volts and still be compliant with the standard. In addition, an ISO standard card could be utilizing either a synchronous, half-duplex block protocol, or a slower byte-oriented protocol and still be compliant with ISO standard conventions and procedures.

Specifications, however, define a much tighter set of rules and operation boundaries for smart cards.

Specifications

The smart card specifications that are emerging within the various vertical industries are quite strict in their interpretation of certain parts of the ISO standards. The first and perhaps most significant of the specifications that have been developed is the EMV specification. EMV is an acronym for Europay, MasterCard, and Visa, the three global card franchise associations representing the 22,000 financial institutions that issue credit cards around the world. These institutions will begin delivering EMV-compliant chip cards for use by their customers by 1998.

The development of the EMV specification followed a series of evolutionary steps. Once the associations decided that a specification was required to facilitate smart card acceptance by consumers, they founded a group to draft the specifications. EMV-1, describing the card and chip environment, was issued within six months. EMV-2, which described the terminal environment, quickly followed. (At this point, the industry began to use the term *terminal* to differentiate emerging hardware from the old point-of-sale magnetic stripe readers.) The most recent version, EMV-3, released in June 1996, describes the specific manner in which cards and terminals will authenticate each other, identify the applications on the card, and exchange data between each device. EMV-3 is one of the more important specifications as other industries (such as the airline industry group, IATA) that use transaction cards for processing consumer applications will tend to follow this general outline and framework for internal databases and data structures, communications between cards and terminals, and overall security controls and procedures.

It is important to note that the estimated 12 million transaction terminals installed worldwide will have to comply with the EMV specification by early in the next decade. The specification provides the common understanding that will create tremendous economies-of-scale effects for card manufacturers, terminal manufacturers, and application developers as new systems are implemented.

TIP

Terminal software should be used to accommodate changes in international standards and industry specifications once a system is implemented and cards are deployed.

Another important specification relates to the telecommunications field. This specification defines the global system for mobile communications (GSM), which includes digital cellular and digital personal communications (PCS) and operates in the 1.8- or 1.9-gigahertz bandwidth. (See Table 3.8.) A specification was developed in the 1980s in Europe through the European Telecommunications Standards Institute (ETSI). This specification, also known as 11.11, is used primarily as the architecture for allowing remote authentication and authorization of digital service within a home region or in a remote area serviced by a company supporting the technology. This specification describes the digital authorization and authentication procedures and programs that are stored in a chip card or a smaller form called a *subscriber identity module* (SIM). The chip card or SIM configures a digital telephone to operate per a described procedure. This SIM-style package can also be used as a security module for many point-of-sale (POS) terminals.

The GSM specification departs from some of the ISO standards in that the handsets and cards do not conform to the standard ISO-style packaging. For example, the GSM specification allows both a 7810-size card and the smaller SIM-size card.

In addition, GSM has attempted to become a more generic specification for the telecommunications industry similar to the way the EMV specification has evolved to become the dominant specification

TABLE 3.8   GSM Features Available

Phase 1
Abbreviated dialing numbers
Short messages
PIN disabling

Phase 2
Charging counter
◆ Price factor and currency
◆ Maximum value
Fixed dialing numbers
Last-number calling
Cell broadcast messages
Language preferences
MSISDN storage

for the financial services world. An important difference between the two is that the EMV specification attempts to reduce specific hard-wired technology factors through the use of objects and generic data-manipulation commands. GSM, on the other hand, still requires very specific hardware interdependency. Unfortunately for GSM, many of the machine-specific instructions deal with the important timing considerations that are required by the technology to establish cellular telephone calls over wireless communications networks.

The development of future specifications will most likely follow the path established by the EMV specification, which depends less on the specific operating environment and the underlying smart card chip and more on the application itself. As we move into the next decade of smart card development, look for more specifications to emerge, many of which may be derivatives of either GSM or EMV (e.g., the Java Card), that will address the requirements of applications using electronic purse/electronic cash cards, stored-value cards, and campus cards in closed or semi-interoperable systems. As electronic commerce validation becomes a requirement for using the Internet, two-way interactive television, or wireless communications, we look for more common ways of communicating.

To support chip cards developed in an environment of changing specifications, the developer must look to the terminal or card reader as the enabling technology. When we look into the future, we expect to find several generations of smart cards coexisting with many generations of terminals. Both technologies expect that a common handshake, speed, voltage requirement, command set, and data interchange will exist to complete a transaction in whatever application is being developed.

No matter how well defined or how lax the standard, the compliance with it becomes the issue.

Compliance with the Standard/Specification

Standards and specifications establish a series of physical requirements that must be met. The only way to demonstrate and ensure compliance is to establish a testing methodology and to conduct repeated and rigorous testing throughout development. This is very important because, given the very small size and form of chip card technology, we as developers are extremely constrained in our ability to provide error checking, exception procedures, and other self-policing technology on the chip cards. Given such a small space, great care needs to be exercised in developing smart card systems and writing smart card programs: Too many or too few parameters for sending subroutine procedure calls, too many arguments or arguments out of processing range may produce unreliable results. This is especially true in some of the cryptographic authentication programs and subroutines available for chip cards. Inaccurate or unreliable results may negatively influence other aspects of the card, since memory, program, and data file protections aren't always available.

When developing a smart card application, testing programs and methodologies should be developed in parallel. These test programs and routines must test both the application and the overall integrity of the card under all possible conditions it may encounter. These conditions include overflow and underflow, illegal arguments being passed in procedure calls, and exceeding file and memory sizes. In some applications the developer must write the memory management routines. Tests must be designed and implemented with particular attention to the management of *cyclic files* (files in which the oldest record is discarded in favor of the most recent record, so that the total number of records remains fixed). Some problems have occurred in the past where the management of cyclical files became unreliable due to bad calculation results as the card underwent the processes of initialization, personalization, and issuance. Care must be taken with design and testing. The same holds true for other types of operations, especially in the security and cryptography area, as these features become increasingly important and more complex as smart card implementation evolves.

The level of effort needed to develop the testing routines is about as great as that required to write the mask and operating system. To ease this formidable task, the acceptance testing of the product may be delegated to a third party or to the customer. In this situation, documentation along with the production programs and test routines will become part of the deliverable work products.

Provability of software code for chip cards is theoretically quite possible. Pieces of code and parts of masks are building blocks that may in time reach an acceptable level of stability, predictability, and provability. When this occurs, the time and level of effort required to verify a smart card application will be reduced. The only areas that cannot be proven mathematically have to do with real-time interactions. Cryptographic applications, such as authenticating a cellular subscriber, will require real-time processing. In dynamic authentication, the best provability will often be what happens in the field.

TIP

It is essential to have very strict design, coding, and testing practices from the first line to the last line when implementing a specification.

In conclusion, standards and specifications evolve. Change is a result of the natural evolution and the maturation of the technology. Typically, cards have a lifetime of two to three years, and terminals have a lifetime of three to five years. The overlaps in life cycles and in generations of technology will not keep pace with the need for speed, memory, and size or the need for additional features. What will be important in the future will be an upward migration path made possible by specifications.

Smart card application development should be as generic as possible. The astute manager will look for ways to develop pieces that will fit modularly with each other to allow growth. While we cannot expect the cards to interact with every type of terminal, terminals of the future will be required to interact with cards developed using multiple specifications, multiple hardware versions of cards, and multiple software levels. Terminal memory sizes and computing power will need to be appropriately sized as well. For a list of important standards and specifications, see Appendix B, Section 2.

CHAPTER FOUR

SECURITY AND CRYPTOGRAPHY

The ability to protect an entire system is one of the advantages of chip card–based approaches over the alternatives. For this reason, a significant amount of effort is invested in improving card system security.

Every chip card system deployed today has a need for security, and security must be a fundamental design component for chip card systems of the future. In addition to improved data integrity, it provides for spin-off benefits, including audit trails and inventory controls.

Security is a constantly evolving discipline. There are no acceptable shortcuts or ultimate solutions. We believe in the philosophy that no single security method, algorithm, key, or procedure is entirely secure. The combination of multiple security components is mandatory to provide a high level of protection against fraud and other threats. The whole, in the case of our model, is indeed greater than the sum of the parts. Existing technologies are being continually improved and new public key algorithms are being announced—for example, the Elliptic Curve Cryptosystem (ECC), aided by a large investment by Motorola, and the RPK algorithm, selected for implementation by the Swiss Federal Institute of Technology.

In this general discussion of chip card security, we consider the following discussion points to be the most important when considering procedures and security management methods for a smart card system design: components of chip card security, hybrid security methods, handshakes and session keys, the sensitive nature of cryptography, terminal security, and the principles of key management.

Components of Chip Card Security

Cryptography is the set of mathematical algorithms used to implement security functions. In our model, security consists of the following:

◆ Data integrity
◆ Authentication
◆ Nonrepudiation
◆ Confidentiality

Data Integrity

Data integrity ensures that the data arrives intact with no tampering or corruption of information. Data integrity is achieved electronically through the application of cryptographic check digits (one-way hashes and message authentication codes) on the data. The hashing procedure produces a small value that uniquely represents the data, much like a fingerprint. If a single bit of data is altered, the hash value will change.

  Data integrity may be achieved by using a Secure Hash Algorithm (SHA-1) as defined by the Federal Information Processing Standard FIPS 180-1. SHA-1 produces a 160-bit hash of the input message and will protect a message or data stream that is less than 2^{64} bits in length. Message Digest 5 (MD5) is another one-way hash function that produces a 128-bit hash of the input message. Authentication is the binding of the sender's (or issuer's) credentials to the data.

Authentication

This process is similar to physically signing a document—a signature is unique and can be recognized later by all parties involved in the transaction.

A digitized signature is a scanned image that can be pasted on any document. It is a numeric value that is created by performing a cryptographic transformation of the hash of the data using the "signer's" private key. A Federal Standard (FIPS 186), defines the Digital Signature Algorithm (DSA) that uses a variable-length key for the transformation of 512 to 1024 bits. RSA, a privately licensed algorithm, on the other hand, uses a key length up to 2048 bits.

The digitally signed data and the digital signature value are transmitted to a recipient. The recipient rehashes the data and then, using the sender's public key, performs a digital signature verification. A successful verification test confirms the identity of the sender. If the test fails, the recipient should be suspicious of the authenticity of the data and the originator of the message. If the test is successful, the message is sent and nonrepudiation becomes an issue.

Nonrepudiation

By definition, nonrepudiation means that a third party can verify the authentication on a transaction and the issuer cannot then deny participation in the transaction. Nonrepudiation will become increasingly important as Internet-based electronic commerce removes the physical presence of a buyer and seller in a transaction.

Confidentiality

Confidentiality is the use of cryptographic techniques to protect information from unauthorized disclosure. The cryptographic approaches of chip card confidentiality fall into two categories: *symmetric* and *asymmetric* algorithms. The key management techniques used to control these algorithms are either static or dynamic.

Symmetric Algorithms

Symmetric algorithms use the same key on both sides of the transaction. This is like going to a locksmith to have two keys made for one padlock. The keys are then given to a sender and a receiver. The padlock is placed on the box used to move a message safely between the sender and the receiver. Because there are only two keys, only the sender and receiver who have the right keys are able to open the padlock and read the message. (See Figure 4.1.)

Static Keys

A static key is a preset key agreed upon by the sender and receiver to secure the desired information. Static keys remain the same for the life of the application or system. Under this type of key management, the keys are established when the cards are personalized. An example of a static key, or algorithm, would be a personal identification number (PIN) authentication where all the security infor-

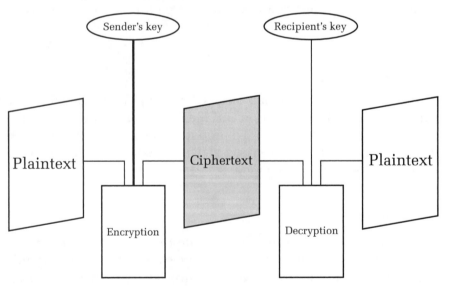

Figure 4.1   Symmetric algorithm encryption and decryption.

mation (key) is precoded and the PIN code is added to this precoded information to create a secure message based on the key and algorithm employed.

Asymmetric Algorithms

Asymmetric algorithms provide a different process to encode and decode a message. In this type of encryption, the key between a sender and receiver is split between a *public* (or known) key component and a *private* (or secret) key component. The secret key component is known only by the sender of the message, whereas the public key is known by both the sender and receiver of the message. A dynamic key is used to manage the asymmetric algorithm.

Dynamic Keys

The use of dynamic keys will become increasingly popular as the worldwide use of smart cards continues to grow. One reason is the ease of updating; key management in very large systems can be a nightmare. Consider the difficulty of keeping all senders and receivers aware of the static key information using conventional means such as the mail or hand delivery via courier for millions of cards!

Public key authorization programs do not require that information be known between all senders and all receivers. The public key can be distributed by mail or posted on electronic bulletin boards. A very high level of security for a large number of cards is possible because the unique private key of each card is different.

The disadvantage, however, is that asymmetric algorithms are mathematically intensive processes and most chip cards today do not have the ability to perform these calculations in a consumer-friendly time frame. Typically, without a special coprocessor on the card, calculation time may be as much as 10 to 15 seconds for even short key lengths. The technology, however, is catching up. In the not-too-distant future, we will see very cost-effective ways to use dynamic keys and asymmetric algorithms in cards and terminals. Many developers are not waiting for these crypto-coprocessor chips to become commercially available and have developed hybrid procedures.

Hybrid Cryptographic Procedures

Developers have employed various permutations of static key management or partial dynamic key management coupled with symmetric algorithms to achieve an acceptable level of security. These hybrid security procedures are the most common procedures used today for financial and other secure transaction processing in smart card applications. This complicated sleight of hand arises because security algorithms push the edge of the envelope of the processing capabilities of the card and the card terminals.

TIP

Dynamic key management and/or asymmetric algorithms must be used to maintain system security.

Architecturally, security systems are very easy to implement. Security systems today use very well understood, well-defined sets of procedure protocols to create secure messages. These include handshaking, which entails a challenge generated by a card to a terminal and a terminal to a card to prove that each is authorized to interact with the other.

Handshakes and Session Keys

This process is typically accomplished through the generation of a pseudorandom number used as a seed. A predetermined calculation is performed using the seed to produce another number. This second number identifies the card. The process is also used in exactly the same way to allow the card to identify the terminal. Handshaking is always bidirectional.

The nice thing about these challenge-and-response-type procedures is that they are not vulnerable to a playback attack. A fraudster initiates a playback attack by electronically trapping and recording the results of a handshake between a card and a terminal. He or she then tries to generate unauthorized transactions and insert them into the system using the recorded results. (This is known in the card world as cryptographic insertion of false messages.) Playback attacks are defeated because each challenge and response calculated alters a setting inside the card and terminal such that when a challenge is repeated using the same seed a different result occurs and the handshake is negated.

One of the outcomes of most handshake processes today is the development of a session key. The session key is a one-time key that is derived and used by a terminal/card reader in a symmetric technology such as DES to encrypt sensitive information. In fact, session key encryption can (and should) be used to secure any information communicated between a card and a card reader.

Session keys for secure transactions regularly use a *triple encryption* or *superalgorithm.* The encryption process encrypts the data. This result is encrypted again with the result passed through the algorithm yet a third time. Triple encryption is a standard practice in the banking industry, as it further complicates the ability to untangle the message information using conventional "lock-picking" attacks. Session keys that are created in a nonrepeatable fashion are dynamic keys. (See Figure 4.2.)

Superencryption is good enough for our industry today. However, as most security professionals will tell you, total security is always an elusive target. Attacks will become increasingly sophisticated as criminals become more cunning in their methods of attacking card systems and their information. Therefore, system developers must continually evolve and improve security methods, which have become so important that they are now a matter of national security. For more detailed descriptions of cryptography methods, see Appendix C.

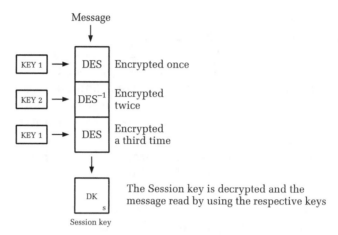

Figure 4.2   Diagram of triple DES encryption. (Courtesy of CP8-TRANSAC)

Cryptography as a Matter of National Security

Developing cryptographic systems is considered a strategically sensitive topic of national security in the United States. The Department of the Treasury's Bureau of Alcohol, Tobacco, and Firearms (ATF) has jurisdiction over the exportation of cryptographic algorithms and procedures. The ATF controls these algorithms by limiting the key lengths to less than 64 bits for the two well-known, commercially available procedures: DES (Data Encryption Standard); and RSA (named after its three inventors). Any smart card system using limited key lengths with these algorithms does not require ATF permission prior to export. All other approaches and longer key lengths will require ATF permission. This has implications on the development of products that may be used not only in the United States, but in other countries as well.

The importing of various chips may be subject not only to U.S. government restrictions, but also country-of-origin restrictions such as those imposed by Germany, France, and the United King-

RSA Implementation

Encryption Process

Alice wants to send a message to Bob; however, she does not want anybody else to read it. She must encrypt the message. To encrypt her message, she runs it through a one-way algorithm that produces a unique value known as the message digest. The message digest is her original message after it has been encrypted and serves as a digital fingerprint for identification purposes.

Before Alice can send her message, she must also encrypt the message digest with her signature key. Together, this produces an encrypted package called the *digital signature.* Before Alice can send the message to Bob, she must also include Bob's public key and her public key for Bob to use to get the message decrypted. These two keys are encrypted with the package, creating the digital envelope.

In order for Bob to decrypt Alice's message, he first decrypts the digital envelope with his private key. This decryption gives him two things: Alice's public key, which was encrypted in the digital signature, and the message digest. Bob uses Alice's public key to decrypt Alice's original message.

At this point, he runs the original message through the same one-way algorithm used by Alice to verify the integrity of the message. If Bob's message digest is identical to the one obtained from Alice's digital signature, then the message was not altered during transmission.

DES Implementation

In a DES system, the same key is used to encrypt and decrypt messages. If a bank or card manufacturer wanted to send messages to cardholders, it would encrypt a message using a key. The message would be transmitted to appropriate parties (e.g., Alice and Bob). Each would be able to decrypt the message using his or her key, which is the same as the bank's key.

DES can be viewed similarly to a lockbox. The owner of the box wants his or her clients to see secret messages. The box is locked and keys are given out to all of the clients. The owner turns the lock one way; to unlock the box, Bob and Alice use their keys to turn the lock back. DES systems are extremely fast in transmitting and unlocking messages. The downside to this technique is lack of key control and a possible overabundance of keys.

dom. These countries have also instituted restrictions and procedures for cross-border transactions.

TIP

Export permission (or waivers) will be required to support cross-border applications.

The ATF may have good cause to worry about security because history shows there are many ways to break a cryptographic system. Devious individuals can try to obtain the keys by deceit, attempt to pick the lock, or guess the way in which information is encoded on the cards or transmitted between the card and the terminal and the data key that is used with that algorithm. The degree of difficulty in cracking a code is directly proportional to the length of the key that is used to secure the data: The longer the key length, the more difficult it is to attempt to pick the lock and decode the data.

Smart cards bring interesting capabilities to the encryption/security war. First, using a smart card that includes embedded processing power offers a more secure way in which to transmit and handle keys in a large-scale environment. When keys are written on a piece of paper or stored on some device such as a personal computer disk, they can be attacked and compromised far more easily than when stored inside a chip card. Chip cards employ a number of techniques to secure and hide keys. One technique is blowing read/write fuses after the keys are loaded on the chip so that even attacks by electronic probes cannot determine the key values. (See Table 4.1.)

Keys can also be loaded during the personalization of the card. Loading keys at this time must be done in a secure facility and under tight controls. Many of the secure card manufacturers certified by Visa and MasterCard will have controls in place in the near

TABLE 4.1 Chip Security Methods

- Chips shall be production-grade quality.
- Sealing coats shall be applied to protect it from environmental or physical damage.
- Opaque tamper-evident coating shall be used to deter direct observation, probing, or manipulation of the surface features of the chip.
- Tamper-detection mechanisms such as cover switches or motion detectors shall be used to detect cutting, drilling, milling, grinding, or dissolving the protective covering.
- Tamper response and zeroization circuitry shall be used to zeroize any plain text keys and other critical unprotected security parameters.
- Environmental failure protection (EFP) shall be used to shut down the module or zeroize the keys if it is outside the module's normal operating range.

Source: NIST-FIP PUBS Processing Standards 140-1, 1994

future to allow them to load keys at the same time they emboss the card and encode the magnetic stripe. (See Figure 4.3.)

By having processing capabilities in the card, keys can also be changed or migrated over time. Key lists can be stored in cards such that more than one key can be issued simultaneously. The addition of various hardware circuits within the chip can achieve cost- and time-effective ways to execute both static and dynamic algorithms.

Since DES and RSA are internationally understood algorithms, they have been implemented by many of the operating system and card manufacturers. Terminal security must also be considered.

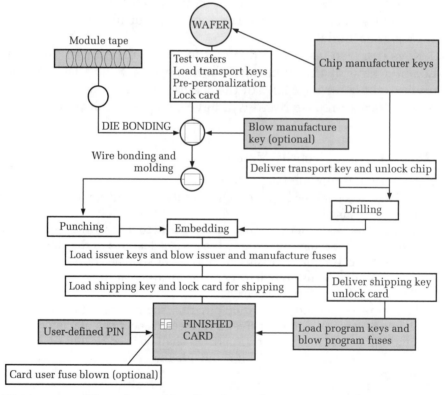

Figure 4.3   Flowchart of loading keys during personalization.

TIP

Always consider buying cryptographic code rather than developing unique security algorithms and procedures.

Terminal Security

In the security world, there is typically a sender and a receiver. In the case of smart card applications, the sender and receiver are often a card and a card-reading terminal. The technologies used inside a card to enable sending and receiving secure messages must be also implemented inside a terminal. Given the growing number of card and security options, it is becoming increasingly important to have flexibility at the reader/writer terminal for the card.

To accomplish this, many manufacturers have adopted an approach that incorporates a security module inside the terminal reader. These security modules are called Security Access Modules (SAMs) or Security Identification Modules (SIMs).

Although various manufacturers have invented a variety of names to label their products, they are all essentially variations of the same smart chip card or cryptographic coprocessor in a slightly different package. In Global Services Mobile (GSM) applications, the security module can be in the form of a small SIM that is inserted into cellular telephone handsets. These SIMs take the form of a small, sealed, plug-in module that contains the required security information used by the application.

In the future, security modules in terminals will become increasingly complex because they will be required to concurrently support a number of different cards, and thus the ability to economically develop applications becomes a complicated engineering challenge. Terminals are physically constrained by the space and slots required to implement security protocols for multiple applications. This limitation will force decisions by the terminal manufacturers concerning which applications to support.

This is an important trade-off in an industry that has to normalize on a small set of different procedures and algorithms so that smart card terminals do not become large and expensive interface devices.

The case study in Chapter 6 will address a typical application of an electronic purse that includes security. The application requires security throughout its entire life cycle from key generation to key distribution to ongoing key management. In the case study, we describe a symmetric DES approach that is accepted by the banking industry for information encryption along with an RSA component used for creating an electronic signature. This signature will verify that the card is the correct terminal for an electronic purse transaction.

Principles of Keys

Security is a very technical field; most smart card developers usually obtain the services of security specialists rather than attempting the programming (and certifying/verifying) themselves. A wide body of knowledge exists, as evidenced in Appendix A, a security reference section listing books and technical security documentation such as the ISO TC67 technical committee on security architectures. Many industries have preferred approaches or methodologies, and the ISO has published several standards as foundation technologies.

However, smart card application developers must have a basic understanding of security. We will forgo a highly technical discussion about security at this point, but you should master the following principals and definitions in order to discuss application security needs with a specialist: key generation, public key/private keys, transport keys, key migration, and key management.

Key Generation

Key generation is usually performed as a centralized function by the card issuer or issuing authority. The actual development of a key set is derived by using the computing power of a host or specialized security computer. Key generation information may be loaded in combination with other specific information on the card (e.g., its serial number, when and where it was made, the manufacturer ID). In the symmetric world, a common master password may reside in one of the levels of keys that are stored in a card.

Public Key/Private Keys

The key generation process is used to place either the public key part or the secret key part on the card, depending on the application. In many instances, the secret key is entered into the cards; however, there is an equally strong argument for storing only the public key half inside a smart card. An issuing key, which is unique to the specific branch or bank location and allows the card to be programmed and issued to a consumer by an authorized entity, is also entered. The number of keys can be quite large, and in some cases a key database and key library need to be designed into the application.

Many of the chip card manufacturers have developed a secure secret area on the card, which can be programmed and then masked, or hidden, to the outside world. The data that is stored in this area cannot be rewritten; this area acts like a one-way door. It can be accessed only via a cryptographic request to process a challenge or to encrypt a message. There are variations by manufacturers on how this is accomplished in hardware and mask software. Typically, there are at least two of these secret areas (registers) available to store keys of various sizes. Over time, the secret area will be enlarged to store a greater number of keys with greater key lengths.

Transport Key

A transport key protects the chip card while it is in transit, such as between manufacturer and issuing authority. The card is effectively locked until it can be issued by a local bank branch or by an issuer that is allowed, for example, to sell you a cellular phone and handset and start your cellular subscription for a telephone. The issuer would have an appropriate key that would be used to open this card and personalize it with specific data. At this point, the key would then change to accommodate the customer and enable the card.

In some instances the issuer can install specific keys and sometimes a key library. These keys are specific to the requirements of the consumer. For example, in electronic purse card applications, the consumer may receive the information for both a credit card and an electronic purse on one card. Thus the keys and procedures for securing the PIN number are stored on the card as well as the procedure for the electronic purse.

Typically, during key generation the keys for both the cards and terminals are derived at the same time. This is important for the design considerations of cards and terminals and for the migration path philosophy of the terminals. For example, some implementations can use a secure methodology to dynamically load keys over a telephone line rather than having to reinstall a physical chip or module inside a terminal. There are several cost advantages to doing this, but there is also some risk of theft or of hackers trying to break the system. These considerations need to be weighed appropriately in the design of the overall system.

Key Migration

What many implementers look for is a periodic migration of keys. Key migration is the process in which one key is used repeatedly for a period of time and then automatically replaced by a new key. In some financial services systems, the algorithms, too, can change over time to keep a thief guessing about what is going on in the overall architecture.

Key Management

Keys today, with only a few exceptions, largely remain static. Once a card is manufactured, it is issued with a set of keys, and in order to change the keys you need to change the card. This is typically the case with GSM cards, pay-TV cards, and many of the financial services and electronic purse cards. Over time, however, we will see increasingly larger numbers of cards that will mirror the way terminals receive keys. As old keys expire and new keys become active, a secure protocol will be required to accommodate the key-refreshing process. Dynamic refreshment of keys and algorithms will make the secret, secure area of the card, loaded during the manufacturing process, increasingly more significant. This area will be used for master keys and algorithms that will allow the dynamic reloading of a card's secured information.

TIP

Security is systemwide, and the smart card is only a component. Cards, by themselves, are not secure.

Security is subject to as much art as science since it will always be constrained by the available space and abilities of the card. How is this done? Most cards today do not have the processing hardware that allows a user to quickly process complex algorithms with long key lengths in real time. Instead, security must be accomplished using software or implementing a partial implementation of a security algorithm. In a partial implementation, many of the results are preloaded and looked up in a table rather than actually computed.

For the next several years, more art than technology will be required to handle multiple algorithms and key lengths greater than 512 or 1024 bits with a microprocessor that has registers of only 8-bit and occasional 16-bit lengths. The successful developer will discover newer ways to break up the processing of keys and algorithms into many different sections and will process that information asynchronously to achieve acceptable time intervals for completing transactions. The advent of affordable coprocessors may allow cryptography to lean more heavily on science, but it will not totally eradicate the need for creativity in solving complex processing problems.

CHAPTER FIVE

Manufacturing the Smart Card

Manufacturing a smart card is a far more complicated and interdependent process than that required for an ordinary magnetic-stripe plastic card. Knowledge of multiple engineering disciplines, specialized card manufacturing, and printing of specialized electronics packaging are all required to produce a finished smart card.

The manufacturing of the card, therefore, is not a trivial process, and you will need to make many decisions for a range of events and interdependent technologies, including card readers, host computers, cards, operating systems, and application software. In this chapter we will discuss the decision-making process in relation to the following: materials used in smart card manufacturing, production alternatives, production trade-offs, the steps required to manufacture a smart card, lead times, and your relationship with the manufacturer.

Materials Used in Smart Card Manufacturing

There are a number of different plastics available for current use, and new compounds are being discovered and brought into commercial use every day. It is important in the design of the application to choose the most appropriate plastic material for application requirements.

The most common plastic compounds, used today in over 90 percent of the industry, are those of the polyvinyl chloride (PVC) family. PVC uses halogen chlorine to stabilize the compound and create many of the properties desirable for a plastic credit card. It is pliable, temperature-resistant, easily embossed with a credit card number and name, and easily printed.

The downside of PVC, as with other halogenated hydrocarbon compounds, is that it has the potential to be a carcinogen. Over time, there has been considerable international effort to move toward more environmentally friendly products. This presents a challenge to the card manufacturing industry, as most of the processes in place today are optimized for production of PVC-based cards.

This shift will not occur overnight, because the investment required to do so is too great. Moving to a different plastic material will change not only the production process, but other steps, such as embossing and inserting chip modules. In addition, equipment used during the issuing of the smart card may also be affected and may need to be changed or upgraded to support an alternative plastic. In the shorter time frame, the industry will increasingly rely on hybrid plastics that use multiple plastic materials to form a card. For example, a card may have three to five layers and use PVC only in the core or center layer. (See Table 5.1.)

TABLE 5.1   Smart Card Materials

Polyvinyl chloride (PVC)
Acrylonitrile butadiene styrene (ABS)
Metacrylonitrile butadiene styrene (MBS)
Polyethylene terephthalate (PETP)
Polybutaline
Polyester PETG
Polypropylene (PP)
Polyethylene (PE)
Polycarbonate (PC)

Is PVC an Environmentally Dangerous Material?

There is considerable debate in the plastics industry on the merits and hazards of many materials in commercial use today. PVC, one of the most common and low-cost plastics available, is used in thousands of products, ranging from medical equipment to household water pipes. At issue is its potential risk to human life and to the environment.

In the early days of PVC, there were considerable problems with the production processes that caused untreated monomers (left over from the polymerization process) to remain with the material. This created a health hazard and did, in fact, cause risk to workers exposed to the material in an unstabilized form. With advances in PVC production methods, this risk has effectively been eliminated, and monomer exposure can occur only when the material is broken down during phases such as incineration, recycling, and other end-of-life methodologies. Due to this risk, there is debate on the prudence and expense of recycling PVC. Alternative methods developed to destroy PVC include chemical processes and very high temperature furnaces.

Are PVC and other highly halogenated hydrocarbons a significant risk to humankind? Yes, given that it is a potentially dangerous material and not easily recycled. In the short term, PVC will be around for many years; however, expect increasingly stronger incentives to move manufacturers toward non-PVC-based alternative materials.

Today in Europe, many larger consumers of PVC-based products are offered financial inducements and other incentives to propose non-PVC plastics to their clients. Alternative plastics include polyester, polybuteline, ABS, and other nonchlorinated plastic compounds. These materials offer sig-

Continued

nificant environmental advantages; however, they are slightly more expensive. One of the reasons that many disposable telephone cards are made from ABS is that they are less harmful to the environment in addition to being recyclable.

Respecting the environment is an important aspect and responsibility of the plastic card industry, although these cards represent only a very small percentage (1 to 2 percent at most) of consumer PVC consumption (by weight). Plastic cards, however, reach almost everyone, and therefore they have a much higher visibility than other, less conspicuous PVC applications such as water piping.

In the future, card manufacturers in the United States will place greater emphasis on environmentally friendly ("green") approaches to plastics production, as is already being done in Europe. While we may never see the end of PVC, we will see an increasingly greater variety of plastic materials and hybrids available in the coming years. Decisions on the types of plastics used for any given application are clearly based on many variables, and consideration of the environment should be one of those variables weighed in the decision-making process.

The raw plastic material comes from a foundry in sheets, rolls, or pellets for injection molding and varies slightly from lot to lot in weight. In addition, the white color of the base or core stock may not be the same shade of white on subsequent orders. These color variations are also possible when stock is procured from different foundries.

Typically, in a printing process there will be expected color shifts based on the chemical interaction between the plastic materials, inks, and other curing or coating compounds. These color differences may be caused by variations in the base color of the plastic material, porosity of the material, and the types of printing ink used, as well as minute chemical variations of the material itself.

When using injection-molded cards (typically ABS-based materials) the percentage of regrind, or recycled waste plastic, reintroduced into subsequently produced cards can vary the flexibility and even the color of the printed, finished cards. As each card manufacturer may use a slightly different process, slight variations in the design and color scheme are expected. In the credit card industry there are variance ranges and acceptable limits established for these color shifts.

Because of these potential variations, not every card will look the same, although they will be pretty close.

More important, if you decide to change materials, say, from a polyvinyl chloride to a polypropylene or a PET polyester, you must understand that there are differences in the physical properties of the plastic as well. Some plastics, for example, will not emboss; they will crack if they are put through an embossing machine. Alternatives to embossing, such as laser engraving used commonly throughout Europe, should be investigated.

Production Alternatives

There are two basic methods for producing smart cards. The most common process is to use existing sheet-offset print methodologies (which produce most of today's conventional plastic cards) and to precisely mill a cavity for the micromodule subassembly (see Figure 5.1). The other option, injection molding, requires the tooling of a very precise card mold, melting the plastic, and extruding it into the mold, forming a card with its cavity already established (see Figure 5.2). Production differences between an injected-molded card and a sheet-offset printed card are noticed in lamination, printing, and milling.

Lamination

Lamination is the outer, glossy, thin (often clear) coating that is common to most credit cards. Laminates are used to protect the card from

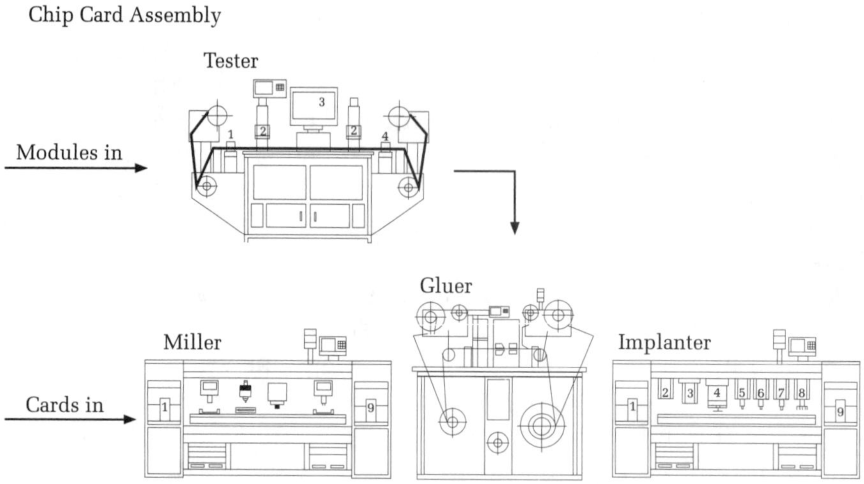

Chip Card Assembly

Figure 5.1   Sheet-fed offset printing process. (Courtesy of Muhlbauer)

excessive scratching and can provide evidence of tampering through inspection. Injection-molded cards are usually not laminated, while printed/milled cards generally are. Since injection-molded cards are produced individually (in a mold), they are typically printed individually as well.

Printing

Printing injection-molded cards requires a very different factory-floor layout than in a typical milled-card plant. Rather than having a few high-speed, offset sheet printers, which can produce many thousands of sheets of cards per day, injection-molded cards use "1-UP" or one-at-a-time printers that individually print the front and back of the cards.

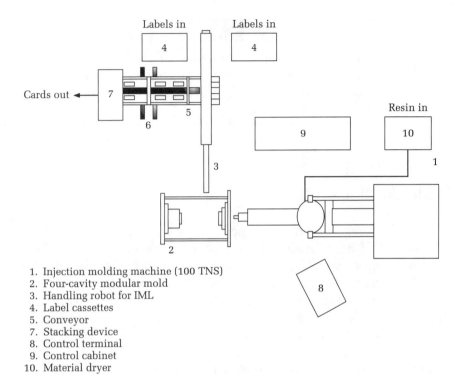

Figure 5.2 Injection-molding process. (Courtesy of GPT Axxicon)

1. Injection molding machine (100 TNS)
2. Four-cavity modular mold
3. Handling robot for IML
4. Label cassettes
5. Conveyor
7. Stacking device
8. Control terminal
9. Control cabinet
10. Material dryer

Milling

Sheet-offset cards must first be individually *blanked,* or cut from their sheet. Blanking produces from 12 to 42 cards per sheet. The blanks are then milled, which creates a small cavity in the plastic to fit the micromodule subassembly. Milling is accomplished at a high-precision, computer-driven, milling station. There are extremely tight tolerances in order to accommodate the smart card chip module. In many production scenarios, this is one of the most common bottlenecks to the process.

TIP

In the future, the pre-personalization of 8K and larger microprocessor cards is expected to become a bottle-neck because of the time required to write the program or data to the card.

Production Trade-Offs

Each of the foregoing processes has production advantages and dis-advantages. When considering which card material will best suit the application, you'll need to weigh a number of trade-offs. For security reasons, features such as holograms, signature panels, and other nonvisible (microprinting) features may need to be included. To provide volume, the sheet-fed offset process may be chosen. It may be necessary to choose lamination for longer life. Other trade-offs to be considered are the pros and cons of milling or molding, run length, card material, card appearance, and card usage.

Milling/Molding

Laminated cards require a milling step, which is an expensive pro-cess, as the card cavity must be precisely formed to accommodate the smart card module. Large variations in milling can lead to cards being rejected (for not meeting the ISO and customer specifica-tions) and to possible delamination of the module from the card cavity because the module does not fit and cannot be adequately glued or affixed in place. The injection-molding process alleviates the need to mill the cavity into the card and produces a very con-sistent (molded) cavity each time. In general, though, the quality of the injection-molded card has, to date, been slightly lower than that of the sheet-offset printed process.

Run Length

Most card factories and card manufacturers are optimized for a certain-size order and card technology (e.g., injection molding). Choosing the right manufacturer may also include understanding its optimal production order size and turnaround times. For instance, smaller printing presses are optimally used for short runs, each for a different image.

Choosing a manufacturer will also include understanding which card material it most often uses and whether that material is appropriate for your needs.

Card Material

Some card manufacturers specialize in particular materials. For instance, ABS is usually the preferred plastic for the construction of injection-molded cards. In the future, hybrid cards made from more than one material may become more common, not only for environmental reasons, but also for their potential extended lifetime and low cost. The card material will, in turn, affect the card's appearance.

Card Appearance

The appearance of the card is an important business decision that is totally independent of the electronic functionality of the card. When considering the type of card material and the processes available, consider how the original artwork may vary from the finalized card product. There are differences in appearance between laminated and nonlaminated cards (i.e., in the glossiness) as well as variations in flexibility and the ability to accept embossing or a photographic image. All of the trade-offs depend heavily on the use of the card, with visual appeal to the customers being an important factor to consider.

Card Use

For many disposable applications, the cost and manufacturing processes slightly favor the injection-molded method. These cards typically do not need to be laminated because their useful life is limited; therefore it's not necessary to use the milled manufacturing method.

The Manufacturing Steps Required to Produce a Smart Card

The manufacturing process has its own intricacies, especially if you decide to split an order between more than one card manufacturer. The manufacturing process and the manufacturers themselves must be orchestrated throughout the following phases to ensure the integrity of the finished product: manufacturing the integrated circuit, placing the integrated circuit into a module, choosing artwork, producing the card, printing the card, embedding the chip module, initialization/personalization of the chip, acceptance testing, activation, and audit reports.

Manufacturing the Integrated Circuit

Volume production of integrated circuits is available only from a limited number of manufacturers. These include Siemens, SGS Thomsen, Philips, NEC, Hitachi, Motorola, and a few others. If your requirements are large, the manufacturing lead time for the silicon must be incorporated into the overall project plan. In addition, the development of the mask and operating system to be burned into the chip may take up to one year to test.

The next step in the process is putting the integrated circuit and the module together.

Placing the Integrated Circuit into a Module

The module is the metal carrier designed to protect the chip and to provide electronic access to the outside world. The chip is wire-bonded to the contacts of the module and then glued to the module itself. The manufacturing process of the module, placing the chip into the module, and testing the assembly can add additional lead time to the project if this manufacturing step is outsourced. Modules are available from only a limited number of manufacturers (Siemens, Philips, and Hitachi, for example).

Another step in the process that will require lead time is the artwork.

Choosing Artwork

Although the artwork may seem trivial, it is probably the most significant reason that cards are rejected. The card manufacturer must satisfy stringent quality requirements that specify color, clarity, positioning, and sharpness. Overlapping all these processes is the production of the card itself.

Producing the Card

As discussed previously, there are two generally used processes for producing a card: the offset sheet process and the injection-mold process. Alternatives in development are expected in the industry in the near future that may greatly improve the economics of mass production of chip cards. As with most technologies, expect continued development and improvement over time. The card is now ready for printing.

Printing the Card

Printing on plastic actually involves management of a chemical reaction. Most inks used in the printing process are carbon-based inks. When the hydrocarbons of the plastic meet the hydrocarbons of the ink, a chemical reaction can take place. This situation can cause color variations, change the levels of receptivity to other inks that may be applied, or alter the ability of the card to accept heat or glue laminate coatings used for antiscratch and antistatic protection.

Furthermore, in many card manufacturing processes, several different layers of materials are brought together in a manner similar to making a sandwich. There are various techniques and technologies used to glue and merge these different layers of the "sandwich" together. Some involve the use of chemicals, while others use heat and pressure. All of these different technologies, including those used for the outer laminate layers, can affect coloration and appearance of the finished card.

Much of the plastic printing industry, therefore, is an art, not a science. There are various proprietary production methods used by the better card manufacturers to anticipate the color shifts that occur. The card seen and approved as a final proof may not be the exact version of the card mass-produced on the assembly line. Therefore, those who are unfamiliar with specifying an outcome for a plastic card need to understand the ways in which the different materials accept inks and the kinds of colors and the levels of resolution (or the level of precision) that will produce satisfactory results. The user must be prepared to specify a low and high tolerance of the color shift according to an agreed-upon color scale.

There will be very little tolerance in the embedding of the card chip module, however.

Embedding the Card Chip Module

The manufacturing step that actually embeds the chip/module assembly in a card is a very precise step. The x- and y-coordinates of the module must be precisely measured before the hole is drilled, and the depth of the cut must be precisely controlled as well.

There are two philosophies on handling this step of embedding the module. The first is to successfully embed a chip into plastic and then print the card; the second is to print and inspect and have only a good, printed product prior to the embedding of the silicon smart card chip. Both approaches have their positive and negative elements. For low-cost, disposable-card applications, such as those used in telephone cards (at 35¢ to 45¢ per unit), it is sometimes cheaper to make the cards and reject them later than to have a careful inspection process of preprinted cards and decide which of those cards will receive chips at the end of the processing step.

An economic assessment between you and the manufacturer should be conducted regarding price and complexity issues and who is assuming the financial risk in the manufacturing process. An industry rule of thumb is to expect to lose between 4 and 5 percent of the starting stock due to manufacturing problems, from either the embedding steps, failure at the testing level, or in rejected art work.

Initialization and personalization are the next steps in the process.

Initialization/Personalization of the Chip

These two processes can be performed by the manufacturer or out-sourced to another company. In both cases, specific information is placed electronically in the chip and mechanically on the card. The information usually found on the magnetic stripe for debit or credit card applications is also written into the chip. The card is then embossed (personalized) with the card number, expiration date, and cardholder's name.

In some cases, pre-personalization, where certain basic data is loaded into the card, can take place at the time of manufacturing. For example, some telephone cards used in prepaid telecom applications such as pay phones are shipped from the factory, loaded, and activated. With this distribution method there are fewer steps required to deliver the cards to the end user or consumer. Other applications such as credit, debit, or electronic purse cards may require that the cards be preloaded with certain base information and security keys, which will require more and more time as the complexity of the application and security increases.

All of this information must then be tested before the manufacturing process can be deemed a success.

Acceptance Testing

Criteria should be developed during the development process for acceptance testing at the end of the manufacturing cycle. However, bear in mind that a certain level of testing of the chip can produce a bottleneck in the whole process. The bottleneck will occur when the manufacturer is not able to simultaneously test a large number of cards. The larger and more complicated cards such as those that contain cryptographic coprocessors can take a minute or more to functionally test from an acceptance perspective. This amount of testing makes the manufacturing/testing time per unit quite expensive. Card manufacturers look at the testing time and the overall effort per piece in establishing their selling price. They also use this information to predict volume and throughput in the delivery of your card order.

A warranty process may also be created and in place by the time the cards are received. If the manufacturer is ISO 9001– or 9002–certified, there should be various levels of certification and accompanying reports. These reports should be documented in detail in your warranty agreement along with the levels of auditability and accountability of the manufacturer. Once the warranty methodology is established, the relationship with the manufacturer will become fairly stable and the customer will be unlikely to change manufacturers. (See Tables 5.2 to 5.4.)

Card activation, the next production step, allows the card to be used within the system.

Activation

Usually, activation is accomplished by a system operator updating the mainframe database. However, your application may be activated by the manufacturer during the personalization/initialization step and shipped to the end user/consumer ready for use.

Finally, a series of reports should be developd to document the process.

Manufacturing/Audit Reports

All of the manufacturing reports must be designed to control each piece of raw material, work-in-process inventories, tested rejects, and finished goods.

TABLE 5.2   ISO 9001/9002 Certificate Requirements

	ISO 9001	ISO 9002
Requirement	x	x
Management responsibility	x	x
Quality system	x	x
Contract review	x	N/A
Design control	x	x
Document control	x	x
Purchasing	x	x
Purchaser-supplied product	x	x
Product ID and traceability	x	x
Process control	x	x
Inspection and testing	x	x
Inspection, measuring, and test equipment	x	x
Inspection and test status	x	x
Control of nonconforming product	x	x
Corrective and preventive action	x	x
Handling, storage, packaging, and delivery	x	x
Quality records	x	x
Internal quality audits	x	x
Training	x	x
Servicing	x	x
Statistical techniques	x	x

T I P

Accountability throughout the process is mandatory. The developer/manufacturer relationship must be such that accountability at each stage of the manufacturing process is clearly established and can be audited.

The audit requirements must be specified in the design along with the acceptance testing requirements. This audit requirement can include conducting surprise and/or scheduled inspections to confirm that every chip, plastic blank, module, security key, and serial number can be accounted for. The audit should also account

TABLE 5.3   U.S. Secure Card Manufacturers/Embossers, Encoders

U.S. Secure Card Manufacturers
Giesecke & Devrient/CardTech
Cardpro Giesecke & Devrient/CardTech
Colorado Plasticard, Inc.
Gemplus
De La Rue Faraday Corp.
Kirk Plastic Co.
ORGA Card Systems
Schlumberger-Malco
NBS Card Services, Inc.
Perfect Plastic Printing Corp.
Perma-Graphics, Inc.
Sillcocks Plastics International, Inc.
Versatile Printing Services, Inc.
William A. Didier & Sons, Inc. DBA Didier Printing
2B System, Inc. DBA Card System, Inc.

Member Embossers, Encoders, and Mail Services
Colorado Plasticard, Inc.
De La Rue Faraday Corp.
Credomatic of Florida
Datacard Corp.
Deluxe Card Services
EFT Source, Inc.
Equifax Card Services
First Data Resources
GE Capital Credit Services
Moore Business Communication Services
Sears National Card Issuance Center
Specialty Network Services, Inc.
Unique Embossing Services
AT&T Universal Card Services
Bank One Financial Card Services
Electronic Data Systems Corp.
First Security Service Company
MBNA Information Services
NBD Service Corp., DBA Computer Communications of America
Total Systems Services, Inc.

TABLE 5.4 Smart Card Manufacturers

Dai Nippon Printing
Data-Pro Card Services
De La Rue Faraday
Gemplus Card International
Giesecke & Devrient GMBH
Interlock AG
Lasercard Systems Corp.
McCorquodale
NBS Smart Card Manufacturing
ODS R. Oldenbourg Datensysteme
Orga Card Systems
Roynix
Schlumberger Smart Cards & Systems
Schlumberger Malco, Inc.
Security Card Systems, Inc.
Sillcocks Plastics International
Telequip corporation
US3 Smart Card Manufacturer

For more detailed information on Smart Card Manufacturers, See Appendix D.

for all manufacturing waste so that any material lost during the manufacturing process is properly disposed of. The audit requirement is especially important when the application specifies that the smart cards will be used as electronic money.

Manufacturing Lead Times

As you consider the design of the overall system, there are lead times involved that are both complicated and interdependent. The first and most obvious are the time requirements for the development of the microprocessor mask, operating system, test routines, personalization routines, and the keys that might be loaded at the time of manufacture. These software programs need to be validated and approved in conjunction with your manufacturer.

The time required to develop and test a mask averages six months for simple applications. As the complexity of the application increases, development time will also increase. As you plan your project, increase the mask development time requirement by three months for every additional feature. For example, to develop a mask for a debit and electronic purse card, the project plan should specify a minimum of 9 to 12 months.

Most smart card developers use an operating system specifically developed for the chip selected for use in the application. The licenses for these operating systems are obtained from companies such as Gemplus, Schlumberger, Orga, or Giesecke & Devrient. Developing an operating system from scratch will add at least an 18-month time requirement to the project. The operating system must be in place and stable well before the mask can be finalized and tested.

The project plan must also accommodate the time required to develop test routines that create all possible combinations of events. Situations such as early withdrawal of the card from the reader or deliberate electronic penetration of the mask must be planned and tested.

As with any complex product, a close working relationship will be required with your smart card manufacturer or manufacturers.

TIP

As you plan your project, increase the mask development time requirement by three months for every additional feature.

Your Relationship with the Smart Card Manufacturer

Specifications are not enough for a manufacturer to create a card and deliver a final product. You should be involved in all of the manufacturing steps described earlier in this chapter to ensure that the finished product meets all of the project's quality criteria.

TIP

It is critically important to establish your expectations in the early stages of the production process. Over time, the day-to-day monitoring requirement will be reduced and only occasional spot checks will be required to ensure a high level of quality.

Once you have chosen your set of manufacturers, you are unlikely to change. Great care should be taken in deciding who you are going to work with to ensure that you are comfortable with the results. Choosing a smart card manufacturer is more akin to creating a partnership than it is to simply placing an order for products in a competitive market. The partnership will involve considerations such as software integrity, security, and preestablished association requirements.

The manufacturing process not only includes the physical creation of the plastic card, but also the application software, test routines, and mainframe code. In the smart card environment, a software bug cannot be tolerated. At today's level of technology, once the program code is burned into the chip, it cannot be fixed if an error is discovered. The success or failure of the project will be determined by the developer's attention to detail at every step of the process.

Many card manufacturers have integrated the mask software development with the card manufacturing process. This integration offers important advantages in supporting the complex security and data personalization needs of a card application. In many cases, constructing the card application based on standard card manufacturer–supplied software masks or kernels can shorten the time-to-market.

Security is also an important consideration in selecting a card manufacturer. For many applications, such as pay-phone disposable cards, each card represents actual monetary value. Therefore, in choosing a supplier, the ability for you to verify the accounting/inventory procedures and audit a manufacturer's facilities are important factors to ensure the quality of material inputs, work-in-process inventories, and finished product outputs.

The credit card associations (Visa and MasterCard) have established requirements that card manufacturers must meet in order to produce association-branded credit cards. These procedures are being revised to reflect the additional steps and requirements of smart card manufacturing. Most of the major card suppliers already have association certification; however, this is not a prerequisite to being able to support smart card production. In fact, there are successful smart card manufacturers in the market today who do not have this certification.

The internal security procedures and requirements that you may need to impose on your suppliers will be based on an internal risk assessment and a review of the current processes in place throughout production and procurement. In the manufacturing of chip cards, it is critical to have 100 percent assurance that the manufacturer has met requirements for each and every step of the production process and accountability of each and every material component.

PART TWO

DESIGN CONSIDERATIONS

CHAPTER SIX

System Implementation Considerations

There are various implementation issues to be addressed before the construction of a smart card system is initiated. The decisions made concerning the system infrastructure will impact the basic design and cost of the chip card. A great deal of thought must be given to the structure and operation of the entire system. The following key design questions must be answered:

- Will the system be fully off-line, an off-line/on-line combination, or fully on-line?
- Will the data managed by the system reside on the card, on the host computer, or both?
- If this is a multiapplication system, will the data reside on multiple computers, in multiple partitions within the (multiapplication) card, or a combination of both?
- What security procedures will be used in the system during the initial launch and throughout its lifetime?
- What provisioning for future security architectures will be part of the initial infrastructure?
- What will be the type of terminals and access methods used for the application(s)?

♦ Will terminals be unattended, attended, or both?

♦ Will third-party terminals be allowed to be installed in the system. Who will certify the terminals with the cards?

Each alternative has unique economic benefits and trade-offs that will impact the overall design of the system. To provide a frame of reference for these questions, we'll view them within the context of an electronic purse case study.

An Electronic Purse Case Study

The objective of this particular electronic purse system is to displace the use of cash for small-value purchases in a university environment. In this application, students will use the chip card in place of cash to buy snacks in vending machines, to pay for meals in the student cafeteria, and to make copies in the library. The application will also facilitate limited electronic commerce by paying for legal, medical, or scientific research conducted on large commercial databases.

The cards will be reloadable. When the value on the card is depleted through use, the card can be recharged at specific loading stations located on the campus. The reload is accomplished by using cash or charging the dollar value to a debit or credit card. When the cardholder graduates or leaves the institution, any remaining electronic value on the card can be reconverted to cash.

This all seems to be a fairly straightforward application. It may seem simple to the user from the front end, but a great deal of planning must go into the back-end design to bring this application to reality. The preceding key questions will be considered in regard to the following issues: card types, amount of memory, data set requirements, security, scripts, failures, information forwarding, testing, terminals, implementation, standards and specifications, alternative technologies, make versus buy decisions, and emulators.

Card Types

For our electronic purse system, we will be using a microprocessor chip card with a defined, secure "purse" architecture programmed into the mask. It is important to address card types and decide which one will be most suitable for the application. We selected this type of card for the pilot for a number of reasons:

◆ Chip cards allow secure two-way communications via the Internet and self-service POS terminals.

◆ The chip itself has the capability to support multiple applications.

◆ The system can be securely managed both off-line (such as for vending machines) and when on-line links or the host computer are not available.

We could have considered other technologies as well, including magnetic stripe cards, memory-only chips, and bar coding. These technologies were rejected for a variety of reasons, among them the following:

◆ Magnetic stripe cards are very vulnerable to fraud.

◆ Memory cards do not afford a high enough level of security.

◆ All of the other approaches require a significant on-line infrastructure that often cannot be economically justified.

We also asked the following question in the selection process: Will more than one kind of card be supported simultaneously? If so, will it be one type of card with different memory sizes or with multiple card types? This decision influences the design and memory size of the terminal and card-reader equipment.

For example, in many international prepaid card telephone systems, the decision to accept rechargeable and EMV-compliant cards (microcontroller-based cards rather than memory-only cards) required a massive reengineering and reinstallation of hundreds of thousands of terminals (public pay phones). Providing for these contingencies during the design phase simplifies the migration path strategy for system enhancements.

To further manage the cost of the system, the designer must also consider the most expensive part of a chip card—the amount of memory required for the application.

Amount of Memory

For our electronic purse application, we will use a 2K or 4K EEP-ROM microprocessor card. The purse will use about 1K of memory, and the remaining space will be available to support other applications.

Smart card semiconductor companies have adopted an approach to allow a single core microprocessor to be linearly expandable in its memory sizes. Therefore, if the system is launched with a very simple (and low-cost) small memory, the same application will be able to run on a higher-memory card, which may also run other programs to support loyalty, credit, and debit applications.

To accomplish this sizing, the designer needs to determine the minimum data set requirements.

Data Set Requirements

Minimum data set requirements will vary by application. Our case study electronic purse requires about 1K of memory, as do most commercially available e-purse systems. In some cases, such as financial services applications, data set requirements may be set by regulation. In a closed-system environment, the designer has to decide, for example, if the card will be anonymous. In an anonymous-card approach, the system doesn't know and doesn't care who the user of the card is. Therefore, memory space does not have to be allocated for a name, an address, or account information.

On the other hand, if the card user is to be identified so that a lost card can be returned or electronically locked to prevent theft, some identification information must be stored on the card. This scenario is just one of the cost/utility trade-offs that the designer must consider.

In Figure 6.1 we have defined a typical data layout showing some of the sizes and data requirements for an electronic purse. As you can see, there are not many data elements required to support an electronic purse application. Therefore the memory requirement is minimized. The data layout and the way in which you will communicate with this data (perhaps with the use of cyclic files) determine what additional information (e.g., transaction histories and balance information) can be stored inside the card. Careful planning of the memory may provide extra room in the card to store loyalty points or buying-behavior data, which may be useful for direct marketing to the cardholder.

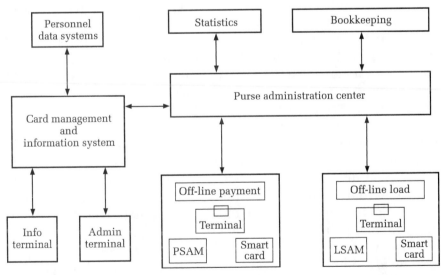

Figure 6.1 An electronic purse data structure. (Courtesy of Siemens Nixdorf)

The cardholder must be assured that these loyalty points and buying history are safe, so security becomes very important.

Security

Our electronic purse operates on the principle that each cardholder has a separate account. The "money" is maintained on the central computer, from both a legal and card-replacement perspective. The card maintains a memo-account value (a copy of the balance on the central computer) as well as the necessary security keys to load (add money to) and unload (make payments from) the card. This purse is therefore *audited,* which means the money is tracked at the host computer as value moves between the cardholder and merchants. The card accounts, however, are encrypted, or scrambled, so there is no way that the system operator can know or link the owner of a specific card to its transactions.

In addition, if the card becomes a multifunction card in the future by loading additional applications, the purse function will remain a separate, stand-alone application.

For financial services transactions that have the capability to update a card (adding to or deducting from an electronic money balance on a card), extra security is required. The most common method used today to implement security is a *secure access module* (known in the industry as a SAM).

Security cards are different from smart cards. While they may share the same architecture and design, their packaging and software differs from that of the chip card. They are usually heavily armored to deter tampering. Security modules are commercially available from a number of manufacturers, such as Siemens, Landis & Gyr, Schlumberger, and Gemplus. These manufacturers offer a security module mask in software that is similar to a card mask.

Security modules, which are dedicated processors for cryptographic algorithms, challenge and response authentication, and key generation, can be treated as quasi-trusted components. Quasi-trusted components have additional (and more expensive) hardware and software protection than do the other components of the system.

However, no single component can be fully trusted. The designer must still implement checks and balances specifically for the security module to ensure that system integrity remains intact.

Security modules have their own masks, operating systems, and communication protocols. The modules differ from manufacturer to manufacturer, and the method in which they are integrated into the hardware and software of the card terminal requires careful analysis. Security module protocols are usually very simple. The smart card initiates a request for a key, message authentication, or a challenge and response. The security module processes the request and issues a result. These black boxes can be implemented in software, and some terminal manufacturers have opted to embed their security modules in the software of the terminal.

The jury is still out on the philosophy that will win the security module game. Clearly, using modules in the form of cards, mini-cards, or plug-in devices offers the advantage of a discreet physical component. If the security software were ever decoded, the modules could be changed out (unlike software-only solutions) before the integrity of the entire system would be compromised.

Security modules have a very reduced capability to store and process information, in contrast to the capability of normal smart cards. This limitation must be taken into account while designing a system that uses security modules. For example, if the system design will allow a number of different cards to be accepted by the terminal, the security module may have an upper constraint of being able to support only two or three different types of cards. In order to accept additional cards, more slots on the terminal circuit board may have to be added to allow for multiple security modules. These decisions must be factored into the system and finalized, as they create additional expense if made later in the implementation.

In our case study, we chose to use a security module that could provide for the processing of a transaction without storing the information inside the security module. Some applications have used the security module as the transaction storage area—which in some instances places stricter limits on the number of transactions and the amount of data that could be stored. Our electronic purse uses data inside the security module to electronically "sign" the transaction data, as well as to update internal counters to ensure that they balance with the host computer.

When designing your solution, considering how all the components will cross-check each other is a critical and significant task. After the system has been designed, it is good practice to request an external audit of the security process to make sure there are no unexpected holes in the system.

TIP

No single security measure will make any system secure; using a mix of methods and technologies will greatly enhance its effectiveness.

Scripts

To update data or regenerate keys, detailed analysis will identify all the various states of the card and terminal during the communications session to ensure correct parameters and data. Event-matrix analysis is necessary to ensure the correct operation of a terminal as well as to detect the presence of a counterfeit device inserted into the system by a third party.

Scripts should allow future commands to address larger files and data fields. The use of parameter-driven tables to describe file sizes and transactions will allow scalable growth onto larger chip card platforms and, in addition, will eliminate the need to use physical address locations in smart card applications. The multiapplication cards of the future will most probably use objects. Careful planning and use of parameter-driven tables will facilitate the introduction and use of commands to manipulate objects. This flexibility is essential for an easy conversion of the application onto a multi-application architecture.

A very important design consideration in the card-to-terminal environment is the possibility of interrupted transactions, or *tearing*.

Failures

An interrupted transaction is a transaction that fails for a variety of reasons, including hardware or software failure in either the card or terminal. An interrupted transaction can also be caused by a cardholder removing the card from the terminal before the terminal has a chance to write the final pieces of information onto the card and close the files.

There are a number of *antitearing* procedures available to protect transactions, the simplest of which is to store status bits inside the card while the transaction is in process. These bits are similar to transaction completion flags used in commercial transaction processing systems (TPS) today. Based on the status of these flags, a torn transaction can be either rolled back and the file returned to the state it was before the transaction was initiated or rolled forward to completion. The disposition of the transaction depends on how a card is used in a terminal.

For example, in a public telephone application, there are two billing methods for telephone service. The system can be designed to deduct the value of the call from the card after the phone call is made. If a tear condition is detected by an open status bit, the next increment or value would be deducted when the card is inserted for a subsequent transaction, and the transaction is forced to completion prior to, for example, displaying the card balance and allowing additional telephone calls.

The second method in our telephone billing example is to roll back the value. Some card-based telephone systems deduct the value from the card before placing the phone call. In this scenario, the card is used to prepay for the unit of time or telephone call and the value is restored to the card if, for example, the phone is never answered or the cardholder pulls the card out of the phone before using the prepaid increment of time. The recovery procedure ensures that the value is returned to the card the next time the card is used.

It is also important that the system record these transaction inter-ruptions to determine if there is an operational problem within the system.

In our electronic purse application, we will handle interrupted transactions with a flag. When set, the flag indicates an interrupted transaction. When a torn card is next inserted into a card reader, it will sense the last transaction was incomplete and react to the situ-ation and condition of the card. The terminal will also send a mes-sage to the host to confirm that the system is in balance before executing the current transaction.

Also important is the detection of a partial failure of either a card or a terminal. Partial physical failures are less of an issue today than they were when card terminals required 21-volt programming voltages. (The terminals had a tendency to "fry" cards.) Partial fail-ure conditions can also be caused by a software defect in either the card or the terminal.

With the introduction of many different kinds of terminals in the marketplace and the implementation of open applications, it is likely that third-party terminals and third-party cards will have to be accommodated by all new systems. Then it will be important to develop exception-handling procedures to provide predictable soft landings in case of a failure or an unexpected error condition. These exception procedures should handle both hardware and software failures.

Hardware exceptions range from the physical detection of an unusable clock frequency to a low-voltage condition. Low-voltage detection is especially important when deploying battery-operated terminals. The test should ensure complete processing. If complete processing is not feasible, it should not be initiated.

TIP

Low-voltage detection is becoming a standard offer-ing by the semiconductor industry for all new chip card designs.

The design process should include an exception-handling script that will reflect the condition and status of both the card and the terminal if, for some reason, a transaction doesn't complete. This could include the setting of tearing bits or other previously described procedures.

Information Forwarding

Forwarding of information upstream or downstream is another design consideration. Clearly, sending errors back to the main terminal is of value to preserve proper operation of the system. In addition to detecting hackers trying to simulate transactions, exception-condition analysis may identify operational problems that may have been introduced into the system by third-party cards and terminals. It is important for the system operator to identify exceptional conditions in order to forestall loss of data integrity and prevent the host database from getting out of sync with the card.

The data design and scripts should include fail-safe procedures that check and cross-check the overall integrity of the system. That is, the design should include the capability for two different events to authenticate each other, which helps ensure that the system and its operating components are working within the design specifications. In the event of a physical failure of any system component, including the cards or terminals, or an incomplete data transmission between a card, terminal, or the host, the transaction could be reconstructed.

Testing

There are various testing techniques, including the use of third-party testing laboratories, that should be used to help verify the system design. For financial services applications, this testing may be required by one of the credit card associations or by a government agency such as the Federal Reserve, Treasury Department, or Comptroller of the Currency in the United States.

After testing, there is the question of certification. Certification proves that the system performs within the design specifications.

Both testing and certification are discussed in greater detail in Chapter 11.

Terminals

The terminals used in our electronic purse applications are standard, off-the-shelf, state-of-the-market products, with the addition of a specific security module as described previously. Purchase transactions are usually conducted off-line. However, transactions can be on-line as well (for computer access or payment, for example). Reloading transactions are handled on-line, except for the cash-to-card stations in which this information may be forwarded off-line.

Transactions are stored in terminals for uploading, which is done automatically on a periodic basis: Each terminal polls the host computer on a predetermined schedule during the nighttime hours. Terminals can upload at other times as well—for instance, when their memory storage becomes full and cannot hold any additional transactions.

After the terminal data is collected, the host computer processes the debits and credits to the card records and makes electronic payments (minus fees) to the merchants. The merchants receive a monthly or quarterly report on the transaction totals (numbers and amounts), which they can then reconcile either manually or electronically (if they receive the report in electronic format).

Any exceptional situations, such as an attempted break-in of the system, illegal cards, or stolen cards/terminals, are reported in an "after-action" report each morning. In an on-line environment, an immediate alert is transmitted (via an electronic pager) to the security officer on duty.

Any smart card application will require a wide variety of terminal products. This includes personal readers such as balance checkers, electronic wallets, and card receipt printers (see Table 6.1). These terminals enable the cardholder to review the transactions and events that have taken place using the card.

TABLE 6.1   Smart Card Personal Terminal Products

Key Fob Balance Reader
Protector Sleeve Balance Reader
Electronic Wallet with Modem Interface
Electronic Wallet
PCMCIA Card Reader/Writer
IC Card Reader/Writer with RS-232C Interface
Mondex IC Card Reader/Writer
Mondex Phone Adapter
Verifone Personal ATM

Source: DAL

Self-service machines constitute a second family of terminal types. Self-service terminals can support a wide variety of applications, including pay phones, parking, vending machines, dispensing machines, copiers, and computer access and services.

Attended machines are the final family of terminal types. They are usually found in a retail environment. A transaction is initiated by a clerk and authorized by the cardholder. In other situations, an attended terminal may be used to resolve card difficulties that require human interaction.

All of these terminal types require different designs based on their functions. For example, a balance-reading terminal may have no buttons at all—you simply insert the card to display the balance.

In the development of a closed electronic purse application, there are several efficient and effective alternatives to implementation.

Implementation

Visa, MasterCard, and Europay association members will be spending millions of dollars to develop and deploy low-cost terminals to replace the 60 to 70 million magnetic stripe terminals in the world and to add possibly hundreds of millions more terminals in locations that currently lack card-accepting devices. Mass production of these devices will provide economies of scale that can benefit the cost-conscious system designer. We would implement our electronic purse system at the university by replacing the most-used terminals first.

Standards and Specifications

An important design element is the analysis of all of the possible communications between a card and card reader. These session transactions and communications should follow a number of preexisting protocols and standards as described in Chapter 3. ISO 7816 (Parts 4 through 7) describes the various commands in detail. Part 5 prescribes that the applications registered on the card be identified according to how the card responds to the reset command. ISO 7816 also provides a generic command set that can be used to move data between a card and a terminal. Some applications, such as GSM, also use preexisting specifications that supersede the ISO standards.

Based on the requirements of the application, another specification to be considered is the EMV specification. EMV has an advantage over GSM in that it allows for the manipulation of objects on the card. *Objects* are data such as name, account number, expiration date, and issuer. Objects may reside anywhere on the card and are not implemented in the rigid hierarchical file structure that is common today. When the data is requested by a terminal-issued command, the data is returned following a number of different paths. In this way, future developers will have flexibility to implement multiapplication cards on which objects may be stored independently of the application.

Determining which standards and specifications will govern development is just a first step in the design. (See Appendix B Section 2 for more information on standards and specifications.) It is then crucial to establish very strict rules governing which of the standard subsets and specification options are incorporated. The level of development effort can be broken into roughly three equal blocks: coding the application, developing the test routines, and creating the documentation. Fortunately, a number of card manufacturers have developed efficient toolboxes and application development platforms that may reduce the total elapsed time needed during the development phase.

TIP

Applications should be developed today with the tightest possible specifications, adhering to the ISO 7816 standard as much as possible. This will allow applications to move across new card architectures as they evolve.

The design decision-making process should also consider alternative technologies.

Alternative Technologies

There are business cases that can be constructed with on-line systems and magnetic stripe cards that are less expensive to implement than off-line smart card systems. The requirements of the system need to be weighed carefully, as a health insurance card differs from a financial services card in its use, type of terminals employed, and required level of security. The application will often dictate the security levels required when information is transmitted between the card and the terminal or host computer. This is especially important to consider when dealing with unattended or self-service terminals that require a significantly higher level of technology, such as the chip card.

Telecommunication-intensive applications, for example, tend to cost more to operate, but they cost less to develop and implement. While developing the overall business case for a system, it is important to consider not only the initial development costs, but also the three- to five-year operating costs of that system. The emergence of new technologies and their financial implications make it important to accommodate future migration strategies. Although this decision may make the system more expensive in years 1 and 2, it may provide the foundation for an even stronger economic return in later periods.

There are trade-offs to consider when choosing between developing a new system or adopting an existing off-the-shelf solution complete with a set of toolboxes to modify it for your specific design needs.

Make versus Buy Decisions

Very often, off-the-shelf products have the advantage of having been field-tested and debugged; however, they often lack the flexibility needed by your specific market requirement. The make versus buy decision tree is very complex. In weighing the alternatives, the designer should assess the organization's capabilities to develop the initial application and provide the ongoing software maintenance and enhancements that most probably will be added during the product's life cycle. Furthermore, should the organization decide to make the product, it must then decide which of the subassembly components (toolboxes and application code pieces) will be purchased and which will be developed from scratch. The following should be considered for purchase:

- Efficient encryption and decryption algorithms
- Challenge-and-response authentication routines
- Data and file management code
- Database management
- Terminal programs

In addition, there are specifications available for prepackaged security components and security modules. These components can be treated very much like black boxes and can be efficiently inserted in the application at low cost.

A final consideration for the design process concerns the use of emulators, as both a design-proving tool and a development aid.

Emulators

As we'll discuss in detail in Chapter 10, emulators have evolved almost to the point where they can be used to build the entire application and test it prior to having the code burned into the ROM. During the design and proving phase, it is usually easier to prototype all parts of the system, including the mask and terminal code.

Designing a smart card application is a very complex undertaking. Even the closed-environment system we used as a reference point in this chapter required hundreds of interdependent decisions. No decision in the design process can be made in isolation, as each one may (and probably will) impact other design decisions. The development team must establish efficient lines of communication to continually trade information about decisions made or contemplated.

CHAPTER SEVEN

FUNCTIONAL DESIGN AND DATA LAYOUT CONSIDERATIONS

Smart card chips are fabricated in a variety of shapes. Most chips are square for stability and ease of handling, whereas a rectangular shape subjects the chip to stress across its longitudinal dimension and increases the chance for breakage. Although chip manufacturers today have the technology to fabricate chips that are 25 square millimeters, when the chip size grows larger than that the risk of lead and semiconductor fracture increases to an unacceptable level.

Chip size will be the overwhelming factor in all design and data layout considerations. In this chapter we will discuss ways to overcome this restraint as it impacts memory, data storage, future enhancements, chip features, file structures, object-oriented structures, and data compression.

Chip Size

Chapter 5 described the chip manufacturing process. An important point to remember is that the chips must fit into cards whose dimensions are determined by ISO 7610. In order to meet these requirements they often have to be reduced in thickness before implantation. The process of thinning, or *backlapping,* the underside of the semiconductor makes the chip brittle, reducing its internal strength, and therefore impacts its degree of reliability. Affixing the chip to a contact plate also requires complex engineering to protect the body of the semiconductor throughout its life.

Because of the size of the chip, there will always be a limited amount of space. This constraint will be with us for the life of this technology, given the small size and footprint of smart cards. Two approaches are being investigated to overcome this constraint: multichip designs and manufacturing chips with higher-density circuits.

Multichip Designs

Multichip solutions are not yet considered feasible by the industry for many reasons, including the potential security threat presented by data communications lines between chips. Also, the manufacturing costs for multichip-based cards is significantly higher than for single-chip implementations.

High-Density Circuits

The use of denser chip structures is the trend of the industry. We are seeing next-generation devices being developed that exceed the performance, capabilities, and densities of all the earlier generations. We expect this trend to continue; however, the physical limitations of silicon will still determine the amount of advancement possible using current fabrication technologies.

Memory

Figure 7.1 illustrates the typical layout of a chip and the amount of space required for each of the different types of storage and processing elements. As you can see, the largest area on a per-bit basis

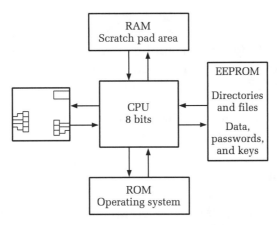

Figure 7.1   Chip card physical layout.

is the RAM, followed by the EEPROM. The smallest area is the PROM or ROM produced during manufacturing.

Semiconductor manufacturers and smart card producers have a number of possible options available to them in designing the sizes of the various elements. Most chip card applications today require a minimum of 8K of program memory to support an application. Many advanced applications such as GSM cards often require additional ROM sizes. EEPROM memory areas provide space for data storage ranging from ½K to 8K or 16K. These larger memory sizes come at a tremendous cost, though, since the physical area of EEPROM memory increases by a factor of 4 over that of ROM.

All card application designers pay careful attention to memory size, making every byte count. Many cards are primarily designed for mass-market applications (i.e., those with very high volume consumer uses). For these applications, minimizing the size of the chip or the semiconductor is critical. Most applications, however, not unlike personal computers, will grow over time. Increased costs and increased capabilities will become a natural progression of the smart card business.

TIP

When designing an application, be aware you may end up supporting a spectrum of various cards running one or more applications at various memory and processing power sizes. Balance the amount of memory required with the maximum amount of capability.

Data Storage

Generally speaking, the more complex a software program, the greater the data storage requirements. For example, many financial applications require very complex security algorithms and keys. The data management of the security procedures alone can occupy a tremendous amount of information space on a card. Storage considerations are threefold:

- ◆ ROM memory (fixed preprogrammed memory)
- ◆ EEPROM memory (data and possibly some program elements)
- ◆ RAM memory (active memory used for processing)

In smart card designs the ROM sizes are growing to meet the increasingly complex programs that are emerging in the industry. The ROM is programmed during the semiconductor manufacturing process, usually called the *masking* process (hence the term *mask* where the code is fixed into the chip). This is unchangeable, with the possible exception of a few fuses that may be burned to lock in security keys or other sensitive or critical data.

The EEPROM is the most discussed consideration of the chip and most affects the pricing of the chip, which in turn most affects the price of the finished card. As the size of the memory increases, so does the surface area of the chip, which in turn increases costs.

The RAM area is usually a few hundred bytes in size and often becomes a constraint in security and processor-dependent applications. As you design programs, it is important to consider the trade-off between RAM size and program complexity. If your application is storing a significant amount of data, it will become difficult, if not impossible, not to use parts of the EEPROM as a temporary scratch-pad memory storage area. This in turn affects the available storage of data (transaction logs, etc.).

Another important consideration beyond the size allocation of EEPROM memory as scratch-pad storage is the cycle time of this memory. RAM is fast and not a constraint in moving data to and from the processor. EEPROM retrieval and, to a greater degree, EEPROM "write" cycles are the slowest and most time-dependent operations. Therefore it is important to economize on mass use of the EEPROM as a primary RAM substitute in the design and layout of an application.

Future Enhancements

An important goal in program design is to support not only the initial application, but a growth path for application enhancements. In addition, you need only to look at the specifications developed by Visa, MasterCard, and the European Telecommunications Standards Institute (ETSI) to see that the evolution to more modern multiapplication cards and specifications is inevitable. These specifications for multiple financial products and various forms of communications are all based on the premise that existing structures must allow for upward and downward compatibility. To achieve this, options that allow flexibility for growth should be part of the initial design criteria.

How do you accomplish this? There are a number of data tools available today, such as object-oriented programming data dictionaries, universal commands, and hierarchical structures. These tools allow for unique growth in dynamic data layout methodologies to achieve some level of hardware and software independence. The degrees of freedom offered in these advanced concepts do not come free. In fact, the price you pay is in additional ROM and EEPROM space.

Emerging Memory Technologies

Chip developments for smart cards are bringing newer memory and processing solutions into commercial use. These include the use of FLASH memory, shadow RAMs, and FeRAMs.

FLASH memory has been commercially available since the late 1980s and is now reaching maturity. FLASH memory technology is a mix of EPROM and EEPROM technologies. With this technology the EEPROM area on a chip is replaced with FLASH, gaining the additional benefits of smaller feature sizes. The term *flash* is used because the entire memory is erased at one time. The name therefore distinguishes the flash devices from EEPROMs, in which each byte is erased individually.

Shadow RAMs, also called NOVROMs, NVRAMs, or NVS-RAMs, integrate SRAM (static random access memory) and EEPROM technologies on the same chip. This technology is used in applications that require fast read and write times, and also require the memory to be nonvolatile. EEPROMs and FLASH memories are nonvolatile, but the read and erase times are too slow for some applications. In addition, shadow RAMs have low capacities.

Another promising memory technology is ferroelectric RAM, or FeRAM, which combines the high speed of SRAM with the nonvolatility of ROM. FeRAMs meet the system requirement of being able to be reprogrammed at high speed with a low-power requirement. FeRAMs will be used in portable electronic devices such as the smart card personal terminals.

FLASH, shadow RAM, and FeRAM manufacturers are refining their processes, and these technologies hold tremendous promise for the future of the smart card industry. These technologies will affect not only the memory sizes (which will increase due to smaller feature size), but also application flexibility through the ability to dynamically update software. These

Continued

advancements will allow a significant telescoping of production lead times to market.

These new memory technologies offer the ability to completely load and reload sections (if not all) of the smart card software. For example, the system operator will be able to dynamically add or remove an entire application on a multi-function card, update or modify subroutines, change card parameters, and so forth. Further, these memories may simplify the steps required to develop a mask, as the only mask information burned into the chip may be software libraries and a low-level kernel to facilitate the loading of application software programs during the personalization process.

Presently, the industry bottleneck in terms of the lead time from an order to delivery is the chip fabrication and the masking of the fixed ROM programs. There are hundreds of masks being developed around the world. Each developer is competing for limited production capacity and each production run of chips needs to be scheduled and built into the production flow. Today these lead times can run in excess of 20 to 30 weeks in some cases. Therefore, any reduction of time to perform this critical process offers tremendous marketing advantages.

A further benefit of these advanced memory technologies is the possibility that generic or standard semiconductors with a loading kernel including low-level management routines may lead to off-the-shelf components that could be added at various stages in the production process. This capability increases the ability to rapidly launch applications as well as to lower total fabrication costs by offering standard, interchangeable parts to the industry.

Monitoring this process is a dynamic task. As we work through the many issues in the integration of new memory technologies, the late 1990s will see the introduction and commercial use of all these technologies in smart card applications.

In the development of the design and data layout, the trade-offs are actually less of an issue today than they were five years ago. As chip manufacturers continue to construct new and innovative processes to drive down costs and improve the capabilities in the microchips, the ability to offer more features in less space is a practical certainty that you can guarantee in your product design.

What does this mean? Basically, if an application is tight on ROM space today, it will have more space in the next-generation chip regardless of the price drop. The reason for this has to do with the way in which the chip logic and circuit capabilities are designed by the manufacturers. To this end, when you are sizing the program and data layouts, it is not a critical trade-off if you design space or features that go beyond the existing memory constraints. It will be a safe assumption that as applications mature and more features are added, the semiconductors will catch up in ROM size and EEP-ROM data storage ability to provide greater flexibility and growth.

TIP

It is important to evaluate one or more chip architectures to decide which best fits the immediate application and the projected development and evolution of the application on that card.

Clearly, software is becoming more valuable than hardware and therefore the cost of the chip is actually less of an issue today. In fact, chip cost is not a constraining criteria in the decision process. The features available on the chip are the most important.

Chip Features

Some chips offer various features that allow for application optimization. These features include long registers for complex security functions, increased memory and processing power, microprocessors, different instruction sets, contactless interchange, and combination modes of interchange. (See Table 7.1.)

Escrow and Storage Registers

An important feature is a secret key escrow and storage register that may be accessed while in the programming mode but becomes a secure and secret area once made operational. This is a very important feature for electronic purse and data security applications and is one that must be evaluated during an overall design assessment.

There are other areas on smart cards that are inaccessible once programmed. These areas are useful for storing information to verify the manufacturer, issuer, or other derived data required for a laboratory card analysis. These areas are also used to store anticounterfeit information that only the issuer, card manufacturer, or semiconductor producer could access. Counterfeiting will probably become pervasive in chip cards, and these secret areas will become increasingly important and expected in future designs.

TABLE 7.1   Matrix of Chip Features

Feature	Reason
Long register	Complex security functions such as key generation
Coprocessor	Faster security
High-speed communications	Faster up/down load (also needed for contact and contactless communications such as radio and other noncontact communications)
Secret key escrow	Required for electronic purse
Storage register	Required for electronic purse
Anticounterfeit information	Security requirement
Volatile vs. nonvolatile	Memory retention without power

Increased Memory and Processing Abilities

While building data layouts and data maps, the designer must keep in mind the idea of increasing memory, processing abilities, and sizes. Today's chips run in the 8K range, and by the time many applications being designed today are implemented, 24K-plus chips will be commercially available. Processing power will also increase, taking the form of additional commands and instructions and perhaps even increased register sizes, from today's 8-bit architectures to tomorrow's 32-bit architectures.

One primary consideration in any and all application designs undertaken today is the migration path. Migration paths that offer upward mobility as the technology increases are commonplace. In fact, just about every application that has been designed for large markets has taken into account increased memory sizes and increased capabilities such as larger key lengths for security coprocessors.

Microprocessors

As market needs for smart cards grow, so will card capabilities. Current chip designs are based on 8-bit microprocessor architectures. These fall into two general categories: Intel 805X and Motorola 68XX implementations. These two chip architectures offer similar instruction sets and abilities and have rich development tools and resources to construct applications. However, slow processor speeds (ISO 7816 specifies under 3.xx or 4.xx MHz) greatly limit their capabilities, especially for security-based applications. Expect and plan for newer chip designs that will not be restricted in speed. (ISO will catch up to the higher processor speeds in time.) Equally interesting is the expectation that newer architectures with larger register and processor sizes will enter the market. In any long term migration strategy, 16- and 32-bit chips are a real possibility that should be considered. Further, future designs may have somewhat different instruction sets than the smart cards in commercial use today.

Instruction Sets

Cryptocontroller and coprocessor implementations use different instruction sets and command structures to implement the highly specialized logic. Converging parts or all of the coprocessor with the primary processor (or adding additional features to the basic processor) will be considerations for future chips.

Development of reduced instruction set computers (RISC) architectures has been proposed as the next evolutionary step of the chip card market. However, potentially low returns may preclude engineering investments to develop these new specialized chips, and therefore the industry will no doubt use as its primary platforms the standard microprocessors available today.

RISC, however, is a technology that may play an important role in the industry and should be monitored. This technology is especially useful for highly complex types of smart cards and may potentially be deployed as a combination cryptocontroller smart card. Therefore, it is important to keep track of developments in this area. Another important technology consideration is the emerging interest in contactless and combination contact-contactless card implementations.

Contactless Interchange of Information

Currently, the design of contactless cards is constrained significantly by the energy available to power and operate a chip in a RF field. The cards used in contactless applications are highly specialized application-specific integrated circuits (ASICs) operating at very low energy levels. The important factor in contactless cards is the size of the antennas. Antenna size determines the amount of energy that can be induced into the card and limits the card's read/write distance and data rate.

Contactless card applications are usually very simple, ranging from detection (card is sensed by the card reader) to the interchange of a short message that may be updated by a card. Common applications include mass transit, building access, ticketless travel, and environmentally harsh (e.g., a factory floor) environments. The data is usually exchanged at a very slow rate and is limited to a few bytes in most close-proximity card applications. Where a contactless card can be brought directly into very close proximity (e.g., inserted into a card reader), more energy can be induced into the card, which will allow for higher-speed processing and higher data communications rates.

Designs for contactless cards are evolving. As described in Chapter 3, a significant amount of work has already led to ISO standards for contactless card operations. With the exit of AT&T, the market dominator, and other early nonstandard contactless cards, we may see this segment of the industry evolve rapidly in the next few years.

Design considerations for the contactless card include the following:

- Distances and field ranges (write ranges are a fraction of read/detect ranges)
- Collision detection (when multiple cards pass within range of the sensing antenna) and resolution
- Processing speeds and data rates based on the technology being applied
- Security of the transmission of data from possible tapping and/or intercept

Combination Cards

COMBI cards are being proposed by a number of manufacturers to combine both the contact and contactless communications abilities into a single semiconductor. This concept, if realized, offers the ability to provide multiple modes and functionality of applications to both in-motion and fixed applications.

Today's COMBI card technology, however, requires further study. The modes of operation may be significantly different between contact and contactless cards, and the performance characteristics of the features between modes must be defined. For example, in the low-power contactless mode, the processor may operate at a fraction of the contact card speed. Therefore, the ability to conduct complex cryptography at the same speeds may be impossible. Further, some designs physically isolate the two different modes. A very limited data channel between the contact and contactless sides of a card is achieved by allowing shared EEPROM memory cells to be accessed from the both contact and contactless elements.

File Structures

File structures are a continuing debate in the advanced card industry. Current ISO 7816 standards specify a directory-type structure, mandating that a card file layout acknowledge the existence of other application data files and data elements. Newer designs, however, in multi-issuer/multiapplication card scenarios contain file structures that hide files and directories from other applications. This file design is important for applications where competitors may coreside on a single card. Companies such as AT&T pioneered many of these security concepts, and today these file structure concepts can be found in some non-ISO-standard card applications. This is an important multiapplication design question: Is it important to conceal the existence of an application on a card from one provider to another?

There are several means to overcome the privacy issue in an ISO standard application. The first is to encrypt the entire file structure and to issue keys that provide access to selected subsets of the directory structure. The downside of this approach is the extra level of overhead that will be required in order to manage the card system and keys. However, this overhead may not be overly expensive given the value for many of these card applications. For example, look at the concept of multiple merchant loyalty programs coexisting on a single payment card: If Sears and JCPenney decided to offer chip-based loyalty programs on a single card, the program could be managed such that neither of the two companies could gain access to the other's data.

The same concept holds true for the airline industry. These applications, therefore, may require separate keys and key management techniques known only to the loyalty system provider and the cardholder. Further, in the management of these keys, the system designer needs to be very careful to accommodate more than one type of cryptographic or security protocol on the same card.

One of the easiest data types to implement is variable and fixed record sizes. Currently, smart card operating systems (see Chapter 8) offer many predefined file types and data types, including the following:

- Fixed-length records
- Variable-length records
- Cyclical records of defined sizes
- Cyclical records of variable sizes

These four data types offer the basic building blocks for all card data storage applications. Cyclical record files are used for most recent to least recent transaction storage, event activity, or security keys that expire after use or time. These cyclical files are managed in a file definition that specifies the size of the record and the number of occurrences of the record. The fields may or may not be fixed length; however, variable-length fields often have limits set on them. File pointers for each record are managed at the directory level of the file structure.

Most file structures are logically oriented and, in advanced card designs, object-oriented as well. This means that systems are no longer required to calculate physical addresses in memory with an associated number of bits to be retrieved. Data is accessed based on a hierarchical relationship or similar data abstraction.

When laying out these records, especially given that increasingly larger parameter sizes will be forthcoming, the designer should use a file or data layout allocation table technique. This technique is used to set parameters in a master file by identifying the number of transaction records that are stored and the size and type of the various files created. This technique allows the designer to migrate to larger memory sizes with the same software. The overhead incurred with this approach is often only a few bytes of additional data on the card and a few extra lines of processing code in the host or terminal software.

When designing smart cards for financial services applications, expect to use approximately 50 percent of the available memory for security programming and data storage. This rule of thumb is based on current memory and processor sizes. Even as security becomes more sophisticated, the same ratio should remain valid, because larger security algorithms in terms of size and complexity will be accompanied by larger and more complex chips.

Object-Oriented Structures

The industry is moving toward object-oriented methods to allocate and store information in smart cards. The EMV specification, for example, makes extensive use of objects to identify and manipulate data elements on a card. The use of objects will help facilitate the use of larger and more complex programs, as well pave the way forward for multiapplication cards. Some of the common objects defined by the EMV specification are also finding use in non-EMV environments. These include CARDHOLDER NAME and EXPIRATION DATE, for example. (See Table 7.2.)

Data Compression

Most card suppliers today offer routines to compress data. Data compression allows for extending the effective storage capacity of a chip card and is an economical and prudent practice in general. Some types of data cannot be compressed safely—such as security keys, biometric data, and program instructions. Text and numeric information, on the other hand, can achieve very effective compression factors. Some suppliers will offer a kernel, or card ser-

TABLE 7.2   EMV Specifications for Data Elements

Data Element—An item of information. A data element is the smallest piece of information that may be identified by a name, a description of logical content, a format and a coding.

Examples of common data elements:
- Cardholder name
- Application expiration date
- Application effective date
- Language preference
- Authorization code
- Issuer public key certificate
- Transaction date
- Application primary account number
- Cardholder verification method (CVM) list
- Transaction personal identification number data

Courtesy: EMV '96 Integrated Circuit Card Application Specification for Payment Systems Version 3.0 June 30, 1996

vices, subroutine that will help to compress data stored on the card. It is important, however, to know exactly what information is being compressed to help ensure that information is not lost during the reconstitution process.

Smart card file designs will continue to evolve as the industry grows. The advent of multiapplication cards requiring both shared and private data will force another revision to ISO 7816 in the very near future.

CHAPTER EIGHT

Operating Systems and the Card-to-Terminal Interface

The interface between the smart card and the terminal provides the physical and logical access to the data on the card. The application code for this interface resides in both the card and the terminal under the control of the respective operating systems. As applications grow in complexity, the physical space in ROM on the card side of the interface will prohibit further operating system functionality unless new approaches are created to implement the interface. In this chapter we shall review the importance of a secure card operating system and the use of security access modules (SAMs) to create the means for the physical exchange of data and control information between the smart card and the rest of the system.

Operating Systems

Commercial operating systems for smart cards are going through a reengineering effort. First-generation operating systems were extremely limited in functionality as a result of a strict interpretation of the ISO 7816 standards. Many proprietary operating systems were developed by a small number of card and semiconductor manufacturers for specific applications. All of these early operating system developers used a hierarchical file structure with file directories to support a smart card–based service. However, the file directory approach limited the application file structures and the services available on the card.

Hierarchical files, as envisioned by the authors of ISO 7816, were adequate when the card functioned mainly as a data repository organized and maintained by the system-based application. However, the rigidity of hierarchical files will slow the migration of smart cards from a data storage device to a device capable of off-line and on-line transaction processing. Transaction processing requires a secure interaction with a terminal or a host in which data elements are created, organized, and shared by several applications on the card.

The hierarchical file structure for card operating systems is still under consideration by the ISO committee, as evidenced by the draft ISO 7816-7 proposal. ISO 7816-7 specifies that data sharing on the card will be accomplished by using the database concept of *views,* which permits different applications to share a common data structure. New data structures required by an application will be created by using the industry standard *Create-Table* SQL command. However, the implementation of these concepts does not solve the problem of processing data with application-specific security requirements, keys, and algorithms.

Today's business requirements mandate that new operating systems be developed to support multiple applications simultaneously while remaining compliant with ISO 7816. The requirements imposed upon the operating system developer by the Europay, MasterCard, Visa (EMV) specification (from the financial services sector) and the Global Services Mobile (GSM) specification (from the telecommunications sector) are fueling the need for change and creativity. New requirements are emerging daily from such diverse organizations as the International Association of Travel Agents (IATA) as well. Within the last two years, a basic financial application has been combined with e-ticketing for the airlines and a loyalty application for "frequent flyers, drivers, and stayers."

Four new operating system extensions have been examined to support the new application requirements: objects, OS kernels and applets, execution security, and application programming interfaces (APIs). Each of these approaches uses the basic building blocks provided by ISO 7816 as an operating system kernel, adding functionality as needed for sharing information between applications in a secure manner.

Objects

The first approach under consideration is to define applications in which the data elements required for a particular transaction are encapsulated together with the processing steps to be performed on them, including the security processing. These self-contained encapsulations are called *objects*. The process includes the following:

- Defining transactions that include both processes and data
- Increasing security through encapsulation
- Creating unique interfaces to multiple applications to ensure data integrity

The object-oriented approach addresses many of the shortcomings of traditional, hierarchical, database-oriented approaches when dealing with multiple applications interacting with data residing on smart cards.

One of the more important emerging research areas is the study of object- and application-oriented OS kernels.

Operating Systems Kernels and Applets

Ongoing research is focused on the development of an OS kernel that retains the features and functions of ISO 7816 while using small modular application extensions (called "applets") to implement the functionality required by EMV, GSM, and IATA. (See Figure 8.1.) The applications are grouped together based on the nature

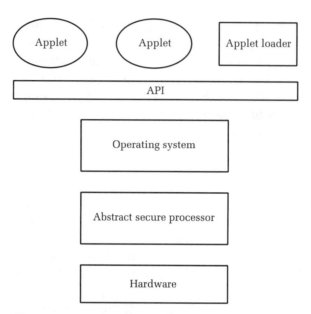

Figure 8.1   Smart card operating system applets. (Courtesy of Integrity Arts)

of the exchange of information between the card and a terminal and the security required to safeguard the transaction. The code required to support each side of the card/terminal interaction is created and maintained in this group or module. Two important features of an applet are increased execution security and the ability to use an application programming interface. (See Figure 8.2.)

Execution Security

The new OS kernels must allow applications provided by different vendors to coexist within the smart card architecture without interfering with each other. Because these applications may not be loaded on the cards at the same time, the OS must provide execution-time security, or a firewall, between the applications and the data to which they have access.

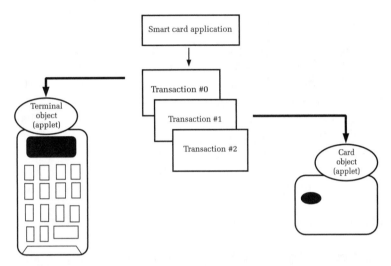

Figure 8.2   Model for smart card applications. (Courtesy of Integrity Arts)

Execution security should include at least the following two features:

◆ A hierarchy of execution privileges (typically, a separate kernel mode reserved for low-level operating system services) and a user mode reserved for application programs

◆ Memory protection routines to prevent an application from extending beyond its allocated memory space into other application spaces or the operating system's space

These features are already present in most main processors intended for workstations and personal computers, and they require only a minimal amount of logic in the CPU. (Memory boundary protection should not be confused with memory management, which often provides physical and virtual data mapping as well.) However, hardware-based memory boundary protection is not present in the smart card microcontrollers that are being manufactured today. Therefore, execution security must be provided by the OS in the form of software.

Application Programming Interface

Supporting value-added applications in smart cards will require a clearly defined application programming interface (API) that will provide a foundation for card-services designers to use in developing application-specific code. Within an object-oriented architecture, the API will be assembled of object interfaces that will support applets. Because memory space is at a premium in smart cards, APIs will likely be optimized for a given type of application. Vertical sectors (e.g., financial services or telephony) would develop and promote their own unique APIs to support their particular applications. An API, or *loader,* is necessary to manage the dynamic loading of applets according to the security policies defined by the card issuers.

We are concerned with the ability to follow standard specifications, especially those for interoperability and for the development of additional functionality once in the field. Even with new card OS developments and the use of applets, it will be difficult to maintain the code, especially after cards are deployed. Once the cards are in the field, they are difficult (and expensive) to recall and modify, unlike operating systems on PCs that can be upgraded by the purchase of new discs or CD ROMs. Once the card operating system is burned into the ROM area of the chip, today's (and the kernel of tomorrow's) operating systems are generally unalterable. In the future, most development activity will be focused not on card operating systems, but on the terminals.

The Card-to-Terminal Interface

Terminal interoperability is far easier to achieve than card interoperability given the greater physical space available. Terminal interoperability is already happening in Europe, driven by the demand for acceptance of different telephone cards in public telephones. The requirement to accept GSM cards from multiple countries (even those with slightly different protocols) and the upcoming field trials for the credit card associations are forcing the development of an open platform. Rather than developing cards and card operating systems that are interoperable with every other card, the strategy of allowing terminals to select and determine the types of cards they will accept (a small universe, which in the future may not be so small) permits faster development and greater flexibility in reacting to evolving market requirements.

TIP

Design the card-to-terminal interface to maximize the potential of the terminal to read more than one card or support only one application.

When designing terminals, or card acceptance devices (CADs), and interfaces, there are two basic elements that must be considered: *communications protocols* and *security*. The companies that have tried to develop universal terminals with interfaces for every type of application have been generally unsuccessful to date. However, as the industry continues to evolve, with only the most robust applications surviving the challenge of the marketplace, it appears that terminals will need to accept only two or three cards to achieve a 90 percent coverage of the major applications. The technologies (and their applications) that communication and security protocols must support include the following:

- Memory type or synchronous communication cards
- ISO 7816–compliant cards
- EMV-type cards
- Electronic purse–type cards

Two issues within the card-to-terminal discussion are broken out for special consideration: communication protocols and security.

Communications Protocols

As applications become more complex, the requirement to transfer larger amounts of data in an acceptable amount of time becomes a compelling design issue. The basic issue with communication protocols is the need for byte- or block- (T=0 or T=1) oriented data transfer. Most of the implementations to date in Europe are T=0 protocol.

A second design consideration is card-to-terminal security.

Security

In these applications the card and terminal will use a challenge and response, or an authentication technique, such that the card can authenticate itself to the terminal and vice versa. Furthermore, some applications require the use of a PIN or a secret code entered by the user as part of the authentication process.

In these applications, the prevalent practice is to also make security modular and shared between the card and the terminal. To accomplish this, the use of a security access module (SAM) is becoming the implementation of choice.

SAMs can be smart cards themselves or smart card chips programmed to reciprocate the authentication process of the card. Another equally valid variation is the use of virtual SAMs that reside inside the logic of the CAD or terminal. These SAMs can be dynamic in that they will accept new algorithms and key pairs while in operation in the field. Most applications to date, however, use static security modules with a very limited capability for field change.

In order to design the card-to-terminal communications dialog and to ensure a high degree of security, a script needs to be developed.

Terminal Script Development

A script is like a play in which the interaction and activities between the actors is predetermined by the author. The ISO standards describe only the very first steps of the card-to-terminal interaction. All that can be expected from a terminal is the reset command and the reset response as described by ISO 7816. Typically, however, interpreting the parameters in the reset reply block will provide information such as the application number, the type and speed of the communications technology, the type of card (synchronous or asynchronous), and the capability of block or byte data exchange during transmission. It is up to the terminal, not the card, to decide how to configure the card-to-terminal session interface. Essentially, after these steps are completed, the designer can define the interaction dialog and procedures of the card and the terminal (unless a very strict and very well defined specification such as EMV or GSM is being implemented). The steps defined by the script depend entirely on the application and determine the requirements placed on the card and the terminal. We suggest the following steps be included in your script: reciprocal authentication, subsequent transactions, transaction confirmation, transaction validation, and transaction failure processing.

Reciprocal Authentication

Using our case study of an electronic purse from Chapter 6, one of the very first steps after the reset command would be the reciprocal authentication of the card and the point-of-sale (POS) terminal. Authentication is accomplished typically through the use of a preprogrammed challenge and response methodology. A pseudorandom number is generated by the terminal and sent to the card with a command to provide authentication. The command can be either the ISO- or EMV-type security command.

The card will return a reply (which may be encrypted) to the terminal using the random number to calculate its authentication information. The terminal will compare the card response to its own calculated result and send a second message to the card acknowledging the authenticity of the card (assuming the calculated results were the same).

After the card authentication process is completed, a reciprocal authentication process will be initiated by the card. The same information can be used. The terminal executes a secure algorithm contained in the SAM and transmits the results back to the card for comparison to the results calculated by the card. Once this mutual authentication has been completed, the next step in the script will depend on the application.

Subsequent Transactions

In the case of an electronic purse, the next step could be a purchase transaction. (This example is for illustrative purposes only. Every application will have its own design approach to implement its script for speed, security, and meeting the business objective.) The authentication processes are the longest steps to complete in terms of time. This is due to the complexity of the cryptographic algorithms being executed in both devices. During the period that authentication takes place, the user is usually prompted to enter additional information into the terminal.

The next step in the process is to complete the actual transaction. In this step a number of application subparts in both the card and the terminal are applied. The terminal will send a command to the card to subtract the transaction value from the balance on the card. Given the privacy considerations of a stored-value application, a card will reply only that it is able or not able to accomplish the transaction. The actual amount of money available to the cardholder will not be displayed to the merchant. For this application, a binary yes or no will be transmitted to specify whether there are sufficient funds inside the card to complete the transaction.

There may also be a confirmation step.

Transaction Confirmation

The cardholder enters a PIN or secret code or presses an affirming Yes button on the POS device to confirm that the amount is correct and that the merchant is authorized to deduct the amount from the card.

The interface between the card and the terminal may or may not need to have a secured channel. When PIN authentication is accomplished, typically it is important that the code or secret password entered by the cardholder be protected from interception by any device other than the card. The authentication result, if issued as a response to an ISO 7816 or EMV command, must be only a yes, a no, or the returning of a certificate (a response from the terminal indicating an acknowledgment that the PIN is correct) that validates the PIN code. As shown in Figure 8.3, the pathway must be secure between the card and the terminal.

Programs that intercept PIN codes, including rogue terminal programs, are a real threat to chip card applications. There have been documented cases of terminals that, for example, copy the magnetic stripe information from bank ATM cards along with the cardholder's PIN numbers. This information is used to produce counterfeit cards that cannot, in many instances, be identified as bogus cards. (Once bogus cards are detected, their numbers will appear on a hot list that is loaded on all the terminals in the system. The next time one of the cards is used, the terminal will kill the card.) While smart cards are inherently more secure than magnetic stripes, protection of the PIN code is an essential design element to forestall theft.

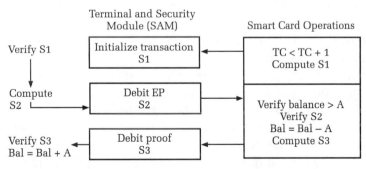

Figure 8.3   The card-to-terminal interface.

Transaction Validation

After completing the interchange of the value, and perhaps the PIN authentication, the card must then process the information and determine if the transaction is valid. The transaction will be valid under the following conditions:

◆ The amount and PIN code are correct.
◆ A verification process approves an account inquiry or a valid request to change transaction.

The transaction can be invalid for a number of reasons, including the following:

◆ Insufficient funds
◆ An incorrect PIN code
◆ Amount incorrect
◆ An unauthorized transaction request
◆ A transaction decline or other message that has been scripted by the application designer

Transaction Failures

The transaction can also fail because of the security keys. The security keys can be either time- or certificate-based depending on the methodology used for securing the transactions. A failure could be a function of an illegal terminal, a terminal out of sync with its host, or a technical fault.

Technical faults should be recorded and addressed as an alert to the system operator. The fault should be analyzed to determine the cause (i.e., electrical failure of the terminal or failure of the card itself). It could also possibly be a software or system design flaw that should be researched.

As described in Chapter 7, a transaction log typically is a cyclical file. However, for fixed transactions and in cases where a limited number of events are authorized by a card, it could be a flat file.

TIP

Normal and exception transactions should be recorded in as many places as possible—for example, in a transaction file in the card, in the terminal transaction log, and, for an on-line scenario, on the host computer via the system operator.

The use of cyclical files is the normal approach used to record receipt logs in the card. These files use a master index for the cyclical file and a pointer record. The pointer record defines the location for the last record in the file and the one just ahead of it. If it is the last available location, the pointer would return to the next logical record that could be used for writing over, which is the oldest transaction in the file.

Cyclical files are useful for recording temporary transaction records, as only the most recent transactions are usually of value. As the time between transactions increases, they become potentially less interesting for the cardholder and for the system operator. When designing these cyclical files, the designer must carefully analyze the amount of data required. Typically, these files consume the largest amount of reusable memory in a smart card, so the trade-offs should be weighed carefully in this type of application.

The designer should also consider how the process of updating various counters and information contained on the smart card will be executed.

Updating Advanced Applications

Advanced applications will require updating at times. Routine updating is useful, for example, to change keys, algorithms, and other control information.

Table 8.1 illustrates some of the commands used in update-type applications. These commands address everything from changing a

TABLE 8.1   Update Commands

Full-duplex transmission block protocol
Change PIN phrase
Delete certificate
Delete key
Generate random number
Get time
Verify signature
Zeroize
Get status
Restore

user-selected PIN code to the updating or deactivating of certain applications.

How to handle expiration dates becomes an important decision in updating applications. We divide our discussion into four parts: a general explanation of expiration dates, multiapplication-card scenarios, loyalty point expiration, and technology obsolescence.

Expiration Dates

There is a very important step in the terminal script that needs to be considered in the development of application cards. This step applies to cards with more than one program or function. Cards are often issued with an expiration date. Expiration dates indicate when a specific card is no longer valid and a new card is required. When the new card is issued, the account number and other fixed information remain the same. Only the expiration date changes. With smart cards, the expiration date could be updated electronically, but many operators have chosen to expire their cards in the same way they expire their magnetic or nonsmart cards today so that they may routinely upgrade the card technology over time. In multiapplication-card scenarios, the issue of varying expiration dates becomes a very important consideration.

Multiapplication-Card Scenarios

For example, if you have a credit and an electronic purse function coresident on the same card, the date of the expiration of the credit card may potentially lock out the electronic purse and the remaining value inside that card. When a card is issued by an issuing institution that is responsible for the primary relationship with the cardholder, the expiration date set by the issuer may be the overriding or master expiration date for a card. In designing multiapplication cards, you need to be sensitive to this point so that the master date expiration does not eliminate the card's ability to read other applications or perform other operations. The advent of multiapplication cards and the reality of bundling more than one application into a card is a situation that requires careful scripting from both the card and terminal perspective.

Loyalty Point Expiration

Another important issue is the combination of loyalty points with transaction cards. Typically, loyalty points have a longer lifetime than a transaction card. Recognizing the coexistence of applications with different expiration dates, the system designer must accommodate the situation in which other applications on the card remain active after a primary application expires. It is incumbent upon the system operator to reissue or replace the card with all the information exactly as it was on the old card or to reset the expiration date electronically when renewing the card.

Technology Obsolescence

The downside of electronic renewal is technology obsolescence. Smart cards and systems are experiencing a massive amount of technology change today. It is in the interest of system operators to replace cards on a fairly frequent basis. Therefore, information storage for the foreseeable future should be on the host and not on the card.

Centralized versus Decentralized Services

Performing decentralized transaction services on the card and centralized transaction storage on the host are activities that must be addressed in the card/terminal script and in the design of your overall system. Many current designs address the issue of storing and processing off-line, user-carried information. We believe that smart cards are a way to provide a more secure authentication between the card and the terminal at the front end with limited information maintained on the card. However, as the global telecommunications infrastructure permits an increasingly on-line society, the need to have an accurate copy of the information maintained on a central database is becoming more and more important. This has to do with whether the card legally contains the information processed by the application or is simply an electronic index card for access to the host computer on which the actual data is stored.

TIP

A centralized data storage approach should be considered to ensure total system integrity.

We recommend that an on-line copy, at least for backup, be an integral part of any smart card system design. The terminal script must be capable of checking the transaction or event counters that are maintained on a card. These event counters accompany the transaction and describe the sequence number or transaction, date/time stamp, or another method of uniquely identifying where and when this card was used. The event counter may also include the latest information about the card balance and the status of the card (e.g., the number of transactions that it has completed in its life).

The event records of all the cards in the system can be transmitted to the host computer so that a reconciliation can occur at the top end. Reconciliation is important for maintaining the overall integrity of the infrastructure. We will discuss the security and fraud-control elements of a card system in Chapter 9.

The card-to-terminal script is a necessary component to protect information and ensure the integrity of the system. The first instance of card-to-terminal interaction occurs when the card is personalized and issued. At card reissuance, when an old card expires and a new card is put into the field, another interaction occurs. At some point in this interaction process, the card must be synchronized with its host computer. This activity will require a well-thought-out terminal script.

Because there are a limited number of operating systems available today, the choice of one over another may dictate the chip manufacturer, security capability, and CAD for the proposed application. Emerging hardware and software technologies and new applications further complicate the decision. A great deal of thought must be given to this topic before a commitment is made to a vendor platform and a specific card or system design, as the decision impacts not only the capabilities of the application, but also the card-to-terminal interface and scripting.

CHAPTER NINE

SYSTEM AND DATA INTEGRITY

The ability to capture and protect information is critical for the safety of any electronic transaction system. System and data integrity must be addressed in all phases of the smart card life cycle—from card issuance to expiration.

In our electronic purse example, internal controls are needed to ensure that the amount of electronic purse money a merchant receives for goods and services exactly equals the amount of the debit to the cardholder's card. To prevent fraud, correct information regarding the status of the entire system must be known so that illegal or "black" money cannot be introduced into the system without being detected.

Data integrity addresses the issue of the soundness of the entire system. We will divide our discussion into the following sections:

- Techniques for providing application protection
- Auditing and security testing
- Trusted third parties
- New security technologies

Techniques for Providing Application Protection

A system's fraud and security process has three objectives:

- ◆ Fraud detection
- ◆ Fraud prevention
- ◆ Deployment of fraud countermeasures to isolate and defeat the fraud

Fraud is one of the main reasons a transaction system, especially one that deals with cash surrogates, must have the protection of host-based audits, verifications, and checks of the system. Despite pretensions to the contrary, no system is totally secure; it is just a question of time and money before a system can be compromised. Intelligent and motivated criminals will find ways, sometimes low-tech ways, to defeat the system.

Several techniques for the protection and integrity of card applications should be incorporated into basic system design: appended security data, transaction logs, unique batch sequence numbers, value-ratio monitoring, systemwide audit controls, off-line procedures, operational rules, secure application modules, and storing value on the host computer.

Appended Security Data

This technique includes checksums (digits) and protocol-error-correcting bits that are appended to each transaction. Message authentication codes (MACs) and other cryptographic signatures transmitted with the transaction information ensure that the information originated from a legitimate terminal and card under control of the system. Sequence numbers and key certificate numbers (or key certificates if they are authenticated) are also appended to a transaction to show which host processing system originated the transaction and how it was originated. Finally, electronic stamps are used to provide date, time, or sequence information to establish a system reference point and prove to the host when a transaction occurred or that the sequence in which it occurred is not inconsistent with other known data.

Another check on the system is the recording of events in the transaction log.

Transaction Logs

One of the most common threats to a system is the nontransmission of information. A transaction occurs but is never processed by the host computer during database synchronization between the point-of-sale terminal and the host.

In this case, the merchant is at risk. Protection for the merchant can be provided through a log that is maintained in a card. The host issues a command for the log to be selectively uploaded if it detects that a merchant terminal has become out of sync with the host computer over a period of time.

In many cases when a transaction is not forwarded it is not an isolated transaction but one of an entire batch of transactions from a particular terminal. These transactions can often be re-created by an exception procedure that reconciles all unsynchronized transactions for a particular merchant number. This procedure identifies which cards may include transaction logs based on past use at that merchant's establishment. They are subsequently queried by the host system the next time they are used.

This approach will recover a high percentage of the nonreconciled transactions. However, not all of the identified cards in the system may be used again during the reconciliation period. Therefore, the reconciliation to that merchant may be less than 100 percent.

An additional design element may include an off-line system. For example, a merchant may log paper receipts from an electronic cash register and use this paper log to re-create and manually post the individual transactions to the host computer.

Unique Batch Sequence Numbers

A second problem occurs when a merchant transmits transactions more than once. To protect the system, the designer should utilize unique batch sequence or serial ID numbers as well as date/time stamps to detect the replay of transactions and avoid the double posting. Employing a technique in which the card and terminal generate unique identification numbers for each transaction when it occurs and for each batch before it is uploaded addresses the threat of a merchant attempt to defraud the system.

The fraud most difficult to detect and prevent is the rogue transaction.

Value-Ratio Monitoring

Rogue transactions, or runaway transactions, are those that are generated by simulation programs. Simulation programs can create hundreds of fraudulent transactions to be presented to the host for processing. This threat has to be carefully evaluated. The best way to do this is to continuously monitor the ratio of the aggregate values spent relative to the aggregate values loaded into the system.

An additional technique is to statistically evaluate any exceptional activity within the card system. The discovery of exceptional conditions may indicate a major failure of the security of the system and the potential that the keys have been corrupted. Most of the commercial systems in operation today have strategies to retire keys and security modules over time. Retiring keys when rogue transactions are detected will help to identify the transactions so the threat can be isolated, attacked, and stopped.

Systemwide Audit Controls

Another threat is transaction modification. This can be difficult to detect in isolated cases; however, using systemwide audit controls like checksums often prevents the threat of a merchant modifying an individual transaction to take more money from a consumer's card than authorized. The system's automated auditor inspects the control registers as aggregate debit and aggregate credit balances maintained on a card and verifies that the values maintained by the host computer are in sync. When the values maintained by the card are not in sync with the host, attempts to change the value of individual transactions may be taking place.

Off-Line Procedures

Smart cards do not have 100 percent reliability. Cards can cease to operate for a number of reasons, including software faults. Similarly, a point-of-interaction terminal, whether it be a home PC, a point-of-sale device, or a vending machine, can also malfunction or cause the card to fail. Thus, transactions may not be forwarded to the host for processing, or incorrect information may be introduced into the system. Because of these multiple points of possible failure, the ability to recover transactions from various locations (card logs, terminal logs, or host databases) within the system or the ability to reconstruct transactions using off-line procedures is very important in the design of the system. On-line audit and control techniques in concert with the store-and-forward abilities of smart cards provide a much higher level of protection than do conventional on-line-transaction systems as they exist today. This design element is not solely for fraud prevention or detection; it also contributes to the maintenance of the overall integrity and auditability of the system.

Operational Rules

In the case of electronic money, it may be important to know the aggregate debit amount in relation to the aggregate credit amount processed by the system in the event that some of the data at a merchant terminal or a card is lost. Operational rules need to be designed into the system to ensure that all transactions are processed in a timely manner. For example:

- Define the maximum period of time from the occurrence and forwarding of a transaction until it will no longer be accepted by a system operator.
- Define the maximum time allowed between transactions before the cardholder must present the card to the issuer for reactivation.
- Establish the value remaining on a card (and refund policy) should it be lost, stolen, or fail to operate.

In many cases, the system operator can dynamically update or replace value on a card by using statistics based on the way the system synchronizes itself and the frequency of synchronization.

Decisions made about data staging and data aggregation require similar design decisions concerning upstream and downstream information transfer. For example, if card balances are to be reset when a lost card is reissued, the system operator may choose to initially restore an approximate value and then provide a reconciliation of that cardholder's account based on transactions made before a specific date at the next reload.

But where does that leave the merchant's compensating balance? For every decision concerning malfunctioning or lost cards, a similar decision must be made to provide the offsetting accounting entry to bring the system into balance. Lost merchant transactions could be represented by a pool of money created when the grand total of debits and credits are reconciled. The system operator may decide to prorate the uncollected or unaccounted pool to the terminals of those merchants who reported failures during a given period.

These policies and procedures have to be designed and implemented as integral to the system. If the system uses on-line terminals such as those utilized by the Internet, synchronization will not be a big issue. In these applications, the card acts as a placeholder and the security token that is exchanged describes the participants on both sides of the transaction. The utilization of an on-line system determines the design of certain types of transactions, just as the use of a store-and-forward system mandates other types of transactions. The choice of system (off-line or on-line) will depend on the application and the value per transaction. In any application, a system using smart cards requires resynchronization in an on-line process to assure overall system integrity.

Secure Application Modules

To provide security and integrity for data, a security module or SAM (secure access module) is commonly used. The SAM is a specific piece of hardware and software in a point-of-interaction terminal that allows for transactions to be authorized off-line and stored for later forwarding, final authentication, and settlement. The system should be designed to anticipate the compromising of a SAM, which again emphasizes the importance of managing the overall system integrity in a fault-redundant way. In some systems, acceptance by the SAM is a "provisional" authorization, which is a business decision by the system operator to accept a possibly fraudulent SAM transaction until the host can actually authenticate and clear the transaction when it is forwarded for settlement.

As different payment and service programs get under way, the need to support multiple SAMs will become increasingly important. To achieve this, many vendors are planning to build terminals with multiple slots to accommodate different SAM solutions. Another alternative may involve developing a single SAM, a "super-SAM," which would accommodate the security requirements of a number of different applications and have the capability to correctly identify and process the various transaction sets at the same time. The economics are clearly in favor of lower-cost hardware. However, a single SAM decision must be carefully considered, as the decision may be difficult and costly to reverse.

Storing Value on the Host Computer

There may be legal ramifications depending on where the value resides in an electronic money or loyalty transaction system.

Financial institutions use audited financial electronic processing systems and computer databases to represent the value of bills and coins. What about smart card systems? Does the value reside inside the card, or do records stored inside the host computer represent the value? Deciding where the value resides is fundamental to the design of an electronic money or loyalty system.

Disposable-card system developers and operators support the philosophy that the card itself—not the host computer—contains the value. Conversely, reloadable system developers, whose cards use more computing power, believe that the actual value should be maintained on the host computer and not on the card itself.

One notable exception to this design philosophy is the Mondex system developed in the United Kingdom by National Westminster Bank, Midland Bank, and British Telecom. The Mondex system is an anonymous, electronic value system in which transfer of value from one card to another card or from a card to an electronic wallet can be accomplished off-line without the intervention of a host computer.

From the legal perspective, we believe that at some point money surrogate or value/reward systems must be audited and reconciled. The majority of system operators support the concept of a central accounting function that balances the system. This is true whether the system is closed or open and whether the value is maintained in one central location or on the card. A central accounting offers the advantage of being able to offer full accountability, auditability, and the ability to restore value when a card fails or is lost or when a terminal malfunctions. Fully anonymous systems, such as the Mondex program, do not offer this capability. The basic design decision must be made within the constraints presented by cost/benefit, security, and operational parameters in conjunction with the regulatory environment in which the system will be used.

Auditing and Security Testing

Periodic auditing practices should be established early in the design phase and executed throughout both the development life

Mondex Security

In a recently published report, it was suggested that Mondex could be compromised by any small commercial engineering firm in the world. In an interview, Ross Anderson, pioneer in the smart card industry and designer of one of the first stored-value card systems, estimates that firms like Chipworks or Intel could break the Mondex scheme, based on the Hitachi S101 and S109 chip, in two to four weeks for about a $100,000 investment.

This claim is part of a larger due-diligence report that was commissioned by the Australian banks to investigate the security of the Mondex scheme. The report covers three significant areas of concern:

◆ Although the chip has been upgraded to the Hitachi S109, the same "significant technical weaknesses could be found in the S109 chip" as in the original S101.

◆ Mondex continues to spread the praise for public key cryptography rather than private key cryptography. While the Mondex system executes public key cryptography in less than two seconds, the argument is moot because Mondex has not disclosed how long the key length is. Performance is directly related to key length in most public key implementations.

◆ The new operating system, MOAS (Multiple Application Operating System), is earmarked for delivery on January 1, 1998 but "is an ambitious project with a high risk of failing to meet the scheduled delivery."

Mondex has many outstanding issues to resolve and is regarded by many as a security risk.

cycle and the operational life of the transaction card system. Security tests must also be conducted as a regular and deliberate practice during the same time period.

It is not enough to test only during program acceptance to see if the system works as designed and documented. To verify that the system properly detects and reports any exceptions or incidents, the implementation of physical dual controls is required to deter insider compromise. In addition, operational procedures that require human intervention need to be continuously verified to assure that when a fraud or an exceptional condition is encountered, the operational staff responds correctly. Data security and system integrity are continuously changing as smart card technology evolves. Some important tests to conduct are as follows:

- Intentionally take the system out-of-balance.
- Red-flag or hot-list specific audit cards and terminals.
- Set conditions to test the alarms and security controls.
- Intentionally alter keys and SAMs.

Trusted Third Parties

Merchant transactions as we know them today are typically accomplished with human-to-human interaction. The Internet and other advanced electronic commerce highways are changing this relationship almost overnight. The transactions of tomorrow will be executed without human intervention and will be accomplished remotely: from the home, an automobile, or while traveling on an airplane.

Remote transactions require a safer and more secure infrastructure than currently exists. Additional data security layers must be implemented in the transactional environment. An additional security layer may be in the form of a trusted third party (TTP) whom a buyer and a seller select to act as a shared, safe authentication provider.

There are several companies seeking to be recognized as TTP providers: MTS Advanced (one of the Canadian telephone companies), GTE Laboratories, the U.S. Postal Service, MasterCard International, and American Express.

Three important aspects of the TTP relationship are as follows:

- The arrangements between the parties
- Entry and exit of the TTPs
- Oversight responsibilities between the parties

The Trusted-Third-Party Arrangement

The trusted-third-party arrangement works in the following manner. One entity (a buyer) desires to initiate a transaction with another entity (a seller) to exchange value for goods or services. Each entity must be registered with the third party, thus establishing the mutual trust relationship. Each entity in the transaction then relies on the trusted third party to authenticate not only itself, but also the transaction between the entities. (See Figure 9.1.)

Authentication of the transaction can be accomplished in several ways:

- Acquire secure certificates from the third party that can be used to encrypt and/or can be appended to messages (such as a MAC).
- Acquire a secure session key that is generated by the trusted third party.
- Derive a key (based on a key received from the trusted third party).
- Include the resubmission of completed certificates after a transaction (at settlement time for example) to authenticate the transaction.

Currently, the commercial use of trusted third parties is very limited. A significant amount of research and development is yet to be

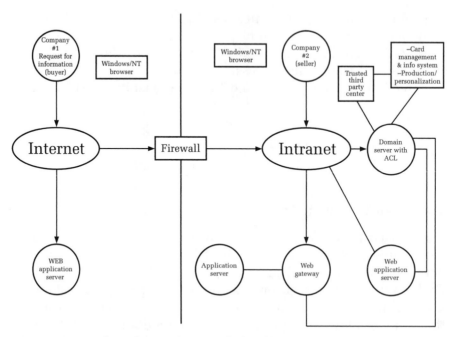

Figure 9.1   Flowchart of trusted-third-party transactions. (Courtesy of Siemens Nixdorf)

done on integrating third parties into the current transactional environment. Visa and MasterCard have developed the Secure Electronic Transaction (SET) specification to enable electronic commerce to use trusted third parties, but we expect the specification will undergo a tremendous evolution during the pilot study and testing.

System developers should also anticipate a significant number of changes in transaction exchange protocols and authentication methods as the TTP concept becomes more widely adopted. Variances in algorithms and encryption methodologies will occur when more than one TTP begins operation in the secure remote environment. In this new environment, the types and processing characteristics of cards, card software, and reader software may change significantly.

TIP

A single card is limited in the number of applications that can be simultaneously supported in the multiple SAM/single SAM environment and with multiple TTPs.

The minimum card memory required to support a value-exchange store-and-forward system is at least 0.5K. As the system becomes more complex, the memory requirements of the smart cards, SAMs, and other components that make up the system increase. The introduction of each new third-party relationship increases the complexity of the steps (programmatic and otherwise) as well as the number of keys and key management components (one-way function resultants, multiple certificates, and possibly more than one key set for time-dependent/expired key considerations).

Today's smart card transaction systems (including most electronic purse systems) have an implicit trusted third party. Specifically, this is the system operator, who processes, clears, and settles the transactions and assures the overall data and security integrity. When the processing and authentication steps are separated (through TTPs, for example), the issue of liability for the integrity of the system also becomes more complex. You should give special consideration to the management and ability to compartmentalize all the individual participants in the event they are compromised.

In the construction of these systems, you must address the management issues of not only the *entry* of each participant, but also the potential *exit* of these participants.

Entry and Exit of Trusted Third Parties

It will not be practical to have merchants (even the virtual ones) exchanging hardware components (such as SAMs) on a regular basis. Further, each and every SAM must be accounted for, as they are one of the trusted components of an overall system. Therefore, procedures must be developed to facilitate the dynamic updating of

SAMs, terminals, PCs, and all other card readers or subcomponents in the system to accommodate the entries and exits of participants.

If multiple transaction-processing companies become involved in the system, their individual authentication and management protocols must also be incorporated. These additional complexities call for the design of easy entry (and exit) for new system participants and applications.

One of the proposed methods of handling the entry and exit of parties is to have short-term keys. These keys would automatically expire after a given milestone based, for example, on time, number of transactions, aggregate value of transactions, and/or a combination of these. Therefore a process to continuously refresh keys and certificates would enable the system to remain relatively current.

As these systems become increasingly complex, so, too, will the need to adopt operating protocols to oversee and control them.

Trusted-Third-Party Oversight

Since there will be no single certification authority for these complex systems, arbitration rules will have to be established to accommodate the multiple certifications required for the multiple programs residing on a card and, especially, for the multiple cards that are accepted by multiple terminals.

Ultimately it may become the responsibility of each individual participating in one or more TTP programs to self-regulate the integrity of its individual component. This would be analogous to loading software onto a personal computer. It is up to the individual to decide which software applications are loaded and unloaded on his or her computer. The same self-regulation may also apply to cardholders, SAM issuers, TTPs, and merchants who operate physical and virtual terminals.

New Security Technologies

New technologies being developed to provide for system and data integrity are as follows:

◆ Application programming interfaces (APIs)
◆ Linking layers
◆ One-way functions
◆ Certificates

Application Programming Interfaces

The development of smart card–aware products is rapidly gaining critical mass. The PC/SC Work Group initiated by Microsoft is one of many initiatives that is helping to drive this awareness. (See Table 9.1.) The concept behind the Microsoft structure is to make smart card–aware applications connect via a common applications programming interface (API). (See Figure 9.2.) IBM, Netscape Communications, Oracle's NCI, and Sun Microsystems together are supporting a second, competitive Java API to provide a standard way of verifying users from any NC terminal. The problem is, the two APIs just don't work together.

Linking Layers

Security via these multiple card input sources becomes increasingly more complex to certify and authenticate. Therefore, requiring a smart card to carry data keys and take on an increasing portion of the processing responsibility to secure transactions will be important. In today's POS environment, every manual or electronic input to the payment system is submitted through a known and proven channel (except for fraud situations in which the transactions try to look like they're valid).

TABLE 9.1 PC/SC Initiative Members

Bull CP8 Transac	Schlumberger
Gemplus	Siemens Nixdorf Information Systems
IBM	Sun Microsystems
Hewlett-Packard	Toshiba
Microsoft	Verifone

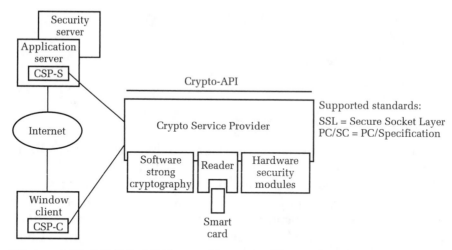

Figure 9.2 PC/SC API layer diagram. (Courtesy of Siemens Nixdorf)

The net effect is an environment that has known inputs. Most of today's electronic purse systems, for example, control the communications between terminals and cards via keys embedded within both the terminal SAMs and the card. Future applications may include devices such as personal computers that do not have keys or SAMs directly attached or embedded inside them.

To accommodate these SAM-less devices, there will need to be an additional linking layer designed and implemented to virtually connect a card and a SAM via a secure transport/communications stream.

One-Way Functions

When exporting this concept to other broadcast media, such as cable and satellite TV, the Internet, and cellular phones, the need for more complex cryptographic systems increases. Today's processing environment, with its limited applications, can support a nominal level of security using static algorithms and rather simple key schemes. However, the use of trusted third parties and different communications technologies requires a different approach to protecting the secret keys in an open environment. One approach to protecting these keys is the use of one-way functions to generate derived keys that can be recalculated by the trusted third party to verify authenticity. Another is the use of secure certificates.

Certificates

Certificates can act as envelopes to carry information between participating entities. The two most popular uses of certificates are for assuring the validity of a certificate (and the identity of the issuer) and for secure transport, where value is embedded within the certificate. (See Figure 9.3.)

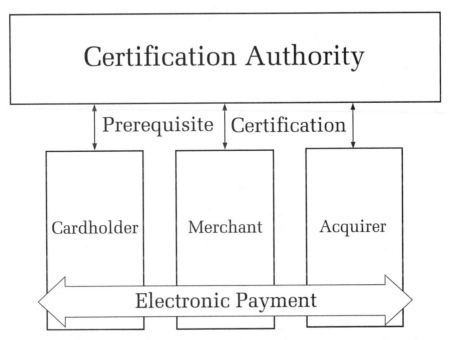

Figure 9.3   Certificate usage in the DigiCash application flowchart.

Security is mandatory to maintain the integrity of the data. When a system is compromised (and it must be assumed that this will happen), procedures must be in place to determine which transactions are valid and which are suspect. One goal of security technology is to help detect and subsequently deter these frauds. The use of multiple checks and balances will help provide data on transactional behavior and the behavior of the SAMs, cards, and the other components of the system. As we move toward multiple-application cards, and especially to multicard-enabled terminals, it will be very important to be able to isolate individual applications, individual transactions, and individual relationships to help detect and cure any possible fraud.

PART THREE

———

IMPLEMENTATION

CHAPTER TEN

SMART CARD DEVELOPMENT SKILLS, METHODS, AND TOOLS

Unique technical skills are required to take a smart card application design and transform it into an executable system. These include the classical data processing abilities such as database design, data manipulation, programming, and so forth. However, distributed processing application development requires the compartmentalization of the physical platforms to facilitate the debugging and testing of the entire system. From a card, terminal, and host perspective, smart card systems are remarkably different and significantly more complex.

Smart card applications require both real-time and time-dependent transactions between devices to implement the security of the system. The development team formed to implement a card-based application needs members who have the same real-time programming skills as those required to build an on-line ATM application. In addition, the team must have the same kind of embedded-system skills required to develop, for example, the main gun-aiming platform of the U.S. Army's battle tank. Embedded systems require tight, error-free code, because the code is burned into the chip sets that comprise the electronics package of the component. Both of these very technical and specialized skill sets are required to fashion a smart card system effectively and in a timely manner. In this chapter we will discuss these skills more fully from the following perspectives:

- The development team
- Development of card software
- Development of terminal software
- Development of host security software
- Testing and certification
- Development and testing tools

The Development Team

Acquiring the resources involved to pull together a chip card–based system is not a trivial task. The programming languages used for chip, terminal, and host application development are different and platform-specific: Assembler and machine languages are used to program first-generation chips; terminal and second-generation chips are programmed in Assembler, C, and C++; and host applications are developed using any number of the higher-level languages such as COBOL or C. Technicians who specialize in only one of these languages are very focused; their skills are often not transferable from one hardware platform to another without an additional investment of time and money for retraining. Technicians who specialize in operating system–level design and programming are very difficult to recruit and retain. In many instances, these technicians may be available only from a card or chip manufacturer such as Gemplus, Schlumberger, or Siemens. Terminal and embedded software developers can be recruited from terminal manufacturers (e.g., Hypercom, IVI, and VeriFone) or from government defense contractors (e.g., General Dynamics, Northrop Grumman, Harris Corporation, and Lockheed Martin). Technical resources at the host level may also pose a recruiting problem depending on the host computer selected for the system. For example, Tandem programming skills are in short supply worldwide today.

The most critical skilled resource, and one that will determine the success or failure of the project, is the security specialist who weaves the tight blanket of physical and logical integrity throughout the system. This individual must have knowledge of modern cryptographic systems such as RSA and DES, key management techniques, semiconductor design, and the ability to apply this knowledge within the multiple physical environments that make up a chip card system. For more detailed information on the development team, see Appendix E.

Once the decision has been made concerning which card technologies to use, potential methods for training the technical staff are vendor-sponsored projects and vendor-supplied courses on card-level programming. In many instances, there will be opportunities to split the work between the vendor and the system developer, allowing the mask-level coding to be accomplished independently by the card manufacturer or facilitated by the manufacturer's programmers and project leaders. In other cases, joint development is a good negotiating point when selecting a potential working partner. Your choice may hinge on the potential partner's capability to dedicate the supplemental critical skills to the project.

The programming required for a chip card system takes place on three distinct hardware platforms:

- The chip card
- The terminals
- The host computer

Chip Card Development

The chip card workload (in terms of time and cost), after completion of the design phase, is roughly divided into three parts: One-third of the effort will be coding the actual software application; one-third will be required for the documentation supporting the programming effort; and one-third will be expended on the test routines to validate the design and the programming. If certification by a third party is envisioned, development time and cost will increase.

Terminal Software Development

In terms of terminal software development, too, the workload is approximately one-third coding, one-third documentation, and one-third testing. About one-half of the coding effort will be dedicated to the development of detailed embedded code. However, the most difficult of these three efforts is the scripting during the testing process to simulate potential inputs, outputs, and consequences of real-time events. These events may include such things as a consumer pushing keys at the wrong time or pulling the card out of the card reader slot at a critical moment.

The host system development effort is similar in many ways.

Host Development

Project management increases somewhat in terms of allocating additional thought and resources toward bidirectional transaction flows and modeling the security and integrity of the overall system. The security-related work elements alone can add one to two full-time personnel simply to explore these issues and implement reasonable and viable solutions. There are security benchmarks available, such as the Visa-MasterCard Secure Electronic Transaction (SET) specification, that provide a baseline for issuing certificates. All components of a secure architecture design, even if they have been used within a closed system, must still be examined and refined (and improved) for operating in an open environment.

Developing card systems today is just as complicated as developing large, integrated application systems was 10 years ago. The fundamental skill sets of project managers, designers, analysts, security specialists, and programmers are required in a team effort. Communications, a key element in the past, is even more important in an open system card project. Each development team at the host, terminal, and card level must interact with the others to ensure that the physical and virtual interface between each component of the system is seamless and watertight.

To ensure that this happens, the project manager and the security specialist(s) must conduct detailed design reviews for the logical interfaces between each component of the system.

TIP

As the development effort progresses, you must test and script your interfaces physically. The necessity to schedule detailed coordination meetings cannot be overemphasized, nor can their value be underestimated.

Development of Card Software

Development of card software today is essentially a process of assembling subroutines obtained from card manufacturers or other software developers and perhaps customizing them to meet the design objective. We are still at least one generation away from object-oriented programming for cards and terminals. The use of card applets, which are object-oriented structures inside the cards, and the development of flash memory will greatly simplify the programming process for cards, eliminating the long lead times (often over 50 weeks from start to finish) currently needed to develop and implement mask software. An even faster development cycle will be possible by using one-time-use programmable logic on top of applets.

When evaluating commercially available card-level subroutine and utility software, an important feature to check is the number and length of parameter variables that are sent to subroutines. In many cases, base modules do not include parameter-checking subroutines. In the event that a program tries to process a parameter with one too few or one too many variables or parameters with an unexpected length, the code may behave unpredictably. This is one of the most common problems encountered while developing software for smart cards and has been the cause of more than one disastrous system failure.

Unexpected results can also occur between a card and a terminal. This happens when data or parameters are exchanged using the wrong assumptions because of values left in the registers from the previous operation. Predictability is the key to making these applications work.

TIP

If possible, clear all the registers and develop scripts to actually test the interface as parameters are passed between code that you developed (and control) yourself and code that you acquired from another source.

Code for the first-generation cards was developed using Assembler, with a few subroutines written in C or a subset of C. The use of advanced languages with higher levels of abstraction such as C++ and Java is becoming more and more common with the development of better compilers and cross-compilers. Today, there are even code optimizers available for the chips on smart cards. (See Table 10.1.)

The future will bring a higher level of abstraction when an applet programming language (perhaps the next generation of Java) generates code that will be compatible with multiple vendors' semiconductors. This will enable software to run on more than one semiconductor platform almost transparently.

TIP

Do not use vendor-specific programming extensions to develop applications. In many instances, the language extensions will not work on semiconductor platforms obtained from competitors!

TABLE 10.1   Commercially Available Semiconductor Kits

◆ Siemens Development Kit
    C-Compiler SLE 44/66
    Assembler SLE 22/44/66
    Simulator dScope
    Linker/Locater

◆ Siemens Development Kit for Crypto Controller (SDK CRCC)
    SDK CC
    Preassembler (PRASS)
    Crypto Simulator

◆ Siemens Emulator Tool (SET 44, SET 22/66)
    Emulator
    CMS

Courtesy: Siemens AG-*HL* division

There are some notable exceptions to the use of applets or higher-level programming languages. Within the security area, where timing is everything and every machine cycle is critical, encryption and certificate routines will continue to be programmed in the lowest possible machine language. This approach optimizes the available processing capability, which is limited in chip card semiconductors. This is especially true with multiprocessor environments such as those used for crypto-coprocessing-type applications.

Until very recently, memory space has been a critical constraint in most chip card implementations. Today, however, the larger memory sizes and expanded features that are available have reduced (but not eliminated) this concern. More research is required in the field of memory management of both ROM, which is used to store data tables, application parameters, and executable code, and RAM, which is used to store the results of transactions and calculated keys and to temporarily store and exchange data.

Chip manufacturers are committed to increasing the size of the RAM. Just as more memory is needed in personal computers to run today's sophisticated applications, insufficient memory may again become a major constraint or bottleneck in the processing capability of chip cards. RAM, unfortunately, is one of the largest devices inside any microprocessor. To conserve space, more and more precomputed information, tables, and parameter values are being stored in the ROM areas of a smart card.

TIP

When developing applications, always be aware of the limited memory size constraint. Manage registers and memory with great care to ensure that the application does not run out of register space or waste too many cycles swapping information between the EEP-ROM and the RAM.

In order to achieve effective register and memory management, various card manufacturers have developed software routines to store and recall registers on command. Even as the cycle speeds of chip cards increase, register and memory management is still an issue and must be carefully controlled.

In many instances, an application's success or failure depends on the ability of the card and the terminal to communicate with each other effectively and economically.

Development of Terminal Software

During the development process, virtually no work can be done independently of the specifications and applications of the card-reading terminals. In many prior first-generation implementations that had unclear or ambiguous specifications, the terminals would wipe out (clear) the memories of the cards or actually short-circuit some of the electrical lines inside the chips themselves. To prevent this expensive mistake, writing clear and concise specifications is

The Java Card API

The Java Card API is a standard set of APIs and software classes that will run on any existing smart card. It is ISO 7816-4–compliant. Java Card enables developers to build a variety of applications for smart cards. Advantages are as follows:

◆ Chip-independent
◆ Based on an easy-to-use and widely accepted programming environment
◆ Provide enhanced security
◆ Capable of running in small memory environments

One of the more interesting applications for smart cards is electronic commerce. The Java Card API enhances Java's electronic commerce capabilities on the client side because the user can now keep trusted data (private digital keys, digital certificates, transactional information, and money) on a card in his or her physical wallet. Java Card can be interfaced with the Java Commerce APIs and the Java Wallet to provide a seamless client-side environment for developing electronic commerce applications.

With the Java Card API, Java is now scalable from platforms as small as those fitting in the consumer's physical wallet to those as large as mainframes in corporate data centers. Java applets that run in a smart card can be scaled to execute on larger machines.

The Java Card API is targeted at smart card OEMs and other developers of smart card applications who require the following:

◆ Small footprint
◆ Ubiquitous

Continued

- Standard implementation/industry standard language
- Rapid application development environment to take advantage of smart card capabilities

The Java Card API supports the following:

- Boolean, byte, and short data types
- Object-oriented scope and binding rules
- Flow-of-control statements
- All operators and modifiers
- Unidimensional arrays of supported data types

The Java Card API is a platform for developing EMV applications. It is based on the plugs-and-sockets architecture contained in the EMV specification. Any ISO-7816 terminal can be used for Java Card API–based smart cards. This includes terminals that connect to personal computers, point-of-sale-terminals, ATM machines, and transaction kiosks.

The Java Card API is an open specification and has been endorsed by a wide variety of smart card companies, including Schlumberger, Gemplus, Bull, Hitachi, IBM, Motorola, ORGA, OKI Electric, Mitsubishi Electric, Toshiba, and Visa.

an important element that cannot be ignored during the design and development cycle.

In the situation where cards and terminals are going to be codeveloped, the teams should meet on a regular basis in order to synchronize each other's applications and communications protocols. During design walk-through meetings always document the protocols and the number of bytes of data transmitted between the card and terminal.

An upcoming consideration for terminal developers is the multi-application-enabled or universal terminal. Many different cards, some potentially issued by competing companies such as Visa, MasterCard, and American Express, will have to be processed by a common terminal. EMV and other technical specifications describe the basic level of compatibility and operation of the terminal. However, advances in merchant, credit card company, or issuing-institution-specific applications will demand increased software sophistication and hardware capability. The question becomes one of certifying that the terminal application supports the advanced services offered by the card.

It is unlikely that there will be a central authority to certify each and every kind of terminal used to support evolving competitive card applications. Antitrust and restraint of trade provisions will probably forbid oligopolies for certain kinds of applications, such as loyalty programs. Competitive differentiation will be based on the ability to use increasingly complex applets that will be independent from one another. This means that terminals will have a common base-operating system with different applications running on top of it—similar to having WordPerfect and Word for Windows running on the same PC. To achieve this level of interoperability, terminal technology will go through an increasingly complex evolution that will have an impact on how card systems will be launched.

For example, the terminal must be able to interpret the card's reply to the reset command within the context of the transaction being performed. Based on the interaction between the card and itself, the terminal must determine which of the multiple optional applications of the card or terminal will be invoked for that specific session. This decision process must be done securely.

A great deal of care must be given to the dialog scripting in terms of what data moves between the card and the terminal. At the most basic level, the dialog will follow the ISO 7816 standard and, potentially, one or more of the industry card specifications such as EMV. Abstraction beyond these levels requires increased computational ability. Modeling how systems will need to work commercially will lead to the design of a cost-effective terminal model to support a wide variety of card applications. (See Table 10.2.)

TABLE 10.2   Terminal Testing Toolkits

Gemplus	MPCOS-EMV
Strategic Analysis	EZ Formatter
Orga	DoctorSim
Aspects Software Ltd.	SmarTest
Schlumberger	Smartware

Furthermore, terminals may need to support multiple generations of cards. Chip cards will have varying expiration dates and, as with any technology, will evolve with faster speeds, more throughput, and greater abilities, functions, and features. Until the last card within an older generation expires, all the terminals in the merchant acceptance universe will have to be able to support it. Terminals will have to support not only multiple competitive applications, but multiple versions of operating software. This means that memory and application management within the terminal will not be a simple and easy process. In fact, terminals will need to employ advanced memory management and processing techniques to identify the correct application and the correct version of the card and to invoke the routines to access the security modules that contain the correct keys for the application selected. Taking these issues into consideration, the value proposition for the terminal increases geometrically.

While developing these new terminal models, it is not unrealistic to use a PC-type architecture to construct the application framework. Terminals today, and future designs going forward, will have increasingly more powerful microprocessors and larger memory sizes with the capability to accept dynamic, remote application reloading. Newer terminals have the capacity to determine exactly which version of the software and exactly which keys are required to support a given application. The designer will have to bear in mind the level of interoperability required and factor in these design trade-offs from the beginning of the process.

The generic and universal terminals that are being developed today must be able to support four or five applications that will cover at least 90 percent of all uses. There will be specific instances of closed or proprietary systems that will also have to be supported. However, supporting these systems is far less complex since the developer controls many of the variables. It is the interoperability requirement that makes multiapplication support a difficult implementation challenge.

Development of Host Security Software

The final area of development takes a systemwide view from the host computer level. In many chip card systems, the host computer might be nothing more than a very powerful PC or server. Most of the smart card implementations to date use an off-line, batch host computer to issue, process, and control cards through their life cycle. Some may also support on-line-based services such as those that use the Internet. The communications link between the host computer and the points of interaction (which may be traditional POS devices) are far more complicated in terms of communication protocols and information management than traditional magnetic stripe or nonchip card applications.

The data being exchanged between the computer, the card, and the terminal is important to consider. The data must be carefully managed while it is exchanged between the card and the terminal, staged for transmission, and then communicated to the host for further processing. The same degree of control must be exercised as data is returned to the terminal to update the card. All communications in either an on-line or off-line scenario must be accomplished through the synchronization of security keys and under the control of a security methodology.

Key management involves the bidirectional authentication of every component of the operating environment, including the cards, the terminals, and the host computer. (There have been cases where the host computer has been penetrated in order to capture information illegally.) Security is paramount, even when the terminals may be acting as only pass-through devices to allow data exchange directly between the card and the host.

In any mode of communication, there are additional pieces of information, such as certificates and data other than just a card number and a transaction amount, that go into the implicit authentication or authorization of a transaction. A transaction is not legally recognized until it actually arrives at the host computer in many of the store-and-forward scenarios. Furthermore, the security checks and balances outlined previously require synchronization at all system levels.

Synchronization makes two work steps very important in a host-level development project. The first is the implementation process, which uses classical software and system development methodologies, and the second is the development of the procedures and protocols for the certification of the process. The latter is more important since the scripts must define the detailed specifications and standards under which transactions are proven and tested to allow only valid ones to reach the mainframe system.

These techniques are well understood and will also manage the reverse flow of information to update application data and keys on a card. This process involves far more complex interaction and requires the card to be on-line with the host computer to complete many types of transactional services.

Building these bidirectional flows (classical card and transaction systems have been unidirectional) involves greater use of synchronization and far more complex sets of commands at both the card and the terminal levels. When the host is authenticated to the card in a similar manner as the terminal to the card, the process in an on-line environment could be viewed as a pass-through authentication that simplifies matters. Things get more complicated when operating in an open system environment where a third party might actually be handling the keys and certification of the cards, terminals, and the host computer.

In these cases, the network interdependencies greatly increase. You cannot assume you know which authorities will control the technical environment. Most chip card designs need to be flexible to accommodate a multiplicity of authentication certificates, trusted agents, and the like in these open transactional systems.

Under an open systems model, the host computer's workload increases tremendously to manage the two-way transaction flow and to ensure that the correct sequence of keys and key pairs, especially in an RSA-type asymmetric cryptographic system, are used. In these scenarios, the implementation team must manage both the key sequences and the key repository, as well as the necessary security protocols at the host level. Overall system security is only as strong as the weakest link.

The essential component in the product development plan is a testing and certification process.

Testing and Certification

Card manufacturers, at least the larger ones, have a fairly extensive qualification process, which we'll discuss in Chapter 11. It is important to note here that when test programs are developed to debug new logic for a smart card, the test routines and procedures should be used to validate the new chip as well. There are actually two outputs to this development step: One is the actual software that is on the chip, and the second is the testing procedures.

The same can be said for smart card terminals. Terminal development projects usually utilize a classic systems development methodology. There is more memory available to work with. Terminals look like personal computers that can be easily reloaded with new application software. However, the low-level input-output routines and the ROM programming of these devices require testing with the same level of precision as that of a smart card using an emulator or a simulator. The difference regarding terminals is that it is easier to replace the ROM chip by using plug-ins or by replacing the software program with on-line software downloads than it is to replace the chip cards once they are distributed in the field.

The same level of care used for testing chips needs to be taken when testing point-of-sale terminals. In fact, point-of-sale terminals, since they are carrying the bulk of the workload, have far more complicated programs and require far more complex testing.

When multiple cards are accepted and the terminal becomes a universal terminal, it becomes even more important to get the software correct at the terminal level.

In some cases, terminal software can be modified to compensate for a deficiency in a smart card that might exist in the field. This situation was evident in the Ohio Electronic Benefit Transfer (EBT) program that deployed first-generation smart cards. In the Ohio EBT program, there was an incompatibility in the specification of the card manufacturer that was not detected until terminals manufactured by a third party were purchased for the program. The system performed as specified as long as cards and terminals from the original manufacturer were used. When the original cards were used in the third-party terminals, several registers in the card were overwritten. To correct the problem, the terminal software was modified to read the manufacturer's code on the chip and then execute a separate routine to prevent the problem.

The ability to change terminal software will become more important in the future as terminal work-arounds are used to maintain backward compatibility with the introduction of new masks, chips, and applications. Dassault and VeriFone are using this approach to demonstrate interoperability between the American Express, Visa, and MasterCard's Mondex e-cash applications, which are quite different in their protocols, security, and operational modes.

Control of the specification and the testing against the specification is also important. The development documentation should be detailed enough to enable comparison of the actual performance of a card and terminal against the design. This includes a script of which activities that card should and should not perform.

Software certification is another area of consideration during smart card system development projects. Certification is important for high-security applications, including financial services. Certification can be accomplished either in-house or by third-party laboratories. The use of third-party labs is a routine practice for financial services and electronic transactions businesses. The development methodology should include a step-by-step documentation of the process, including the design specifications, the coding discipline, and the encoding techniques. In fact, third-party laboratories will first study the development methodology before they try to attack the product itself.

Development and Testing Tools

The key to developing good terminal software is using the toolkits and code libraries available from terminal vendors. Each vendor has its own set of routines and preferred ways of implementation. In some cases, companies like VeriFone and Hypercom prefer to develop applications under contract so they can maintain control of the technical environment within the terminal. The terminal manufacturer can quickly accommodate the individual specific tasks such as communications and workload management, which may vary depending on the types of devices used in your implementation.

The use of field-proven building blocks that have good commercial history will be one of the factors in selecting a vendor. Furthermore, the process of sharing the development between the end-user community and the card or semiconductor manufacturer is a decision that should be made early in the process. There are some software engineering companies available that specialize in machine-level code to help prove the design, while others may accept an outsourcing contract to complete the work.

Fortunately, there are many testing tools available, ranging from emulation and simulation software modules to debugging hardware for cards and terminals. There are two kinds of emulators: One is a program validator or simulator and the second is a true in-circuit emulation of the semiconductor. An in-circuit emulator is a replication of every single circuit on a very carefully constructed platform, which allows for parametric as well as functional testing of the semiconductor. The price difference between a simulator and a true in-circuit emulator is on the order of 1 to 75.

A simulator is far less expensive and can be executed on a PC equipped with special software and an RS-232 hardware interface. A true circuit emulator can cost over $35,000 to duplicate the electronics of the integrated circuit including all the debugging tools. In most current chip operating system development projects, the use of a simulator is sufficient, as many of the basic programming applets, subroutines, modules, and libraries available from the manufacturer have already been field-tested.

If the portions of the system that are unknown are unique software elements of the card application, then a full in-circuit emulation is not required. Full in-circuit emulation is required when new chips, chip sets, or architectures are being developed or when a significant coprocessor or new hardware logic is added to a proven basic design. In these cases, simulation is not enough, and a full in-circuit test is required to validate the design. Once a design is proven through emulation, it is then taken to a test silicate, which may be made up of several hundred devices, to submit the electronic platform to a full test and qualification procedure.

Developing smart card systems requires a great deal of discipline from the very beginning. The administration, logistics, and costs of managing mistakes after the cards are distributed are prohibitive. Although it didn't seem so at the time, fixing bugs in a host or PC application was simplicity itself: The code was located in a convenient location; it was accessible; once fixed, it could quickly be tested. In the chip card world, discipline during design, development, documentation, and testing is the only thing that stands between success and failure.

CHAPTER ELEVEN

Smart Card Testing and Certification

Testing includes the review of the technical design and the specifications of component technologies, development methodologies, and system documentation. Certification includes the creation of procedures to exercise each and every physical and logical component of the system.

Testing and certifying chip cards and terminals is more complex than the process used for most consumer products. In addition, testing and certification procedures will evolve as developers learn more about how these applications can be used and misused. New and innovative procedures are being developed to certify products at the card, terminal, and host levels.

In this chapter, we will examine the smart card testing and certification process, testing and certification organizations, reducing risk in the testing process, the cost of testing, and the role of engineers in the testing process.

The Testing and Certification Process

A consumer product that can be found in every home and on many street corners is the telephone. Telephone manufacturers had to ensure that their products worked under many different conditions. Over the years, the telecommunications industry created the following three-step model (based on lessons learned, often the hard way, through product failure) to test and certify products before taking them to market:

- Establish test or benchmark criteria
- Test the product against the individual benchmarks
- Test the entire life cycle of the system

Benchmark Criteria

The first procedure in testing and certification is to establish the testing criteria. These test criteria establish the known baseline, or benchmarks, against which the card, terminal, or host system will be measured. Benchmarks should cover all peripheral components to ensure that the equipment operates within the documented specifications and has been implemented to both the letter and intent of the overall design.

Established benchmarks to measure the performance of various components of the system infrastructure are available. The ISO standards and industry specifications provide many of these physical and logical benchmarks, which can be applied at various stages through the design cycle. In some cases, benchmarks may be little more than paper-based flowcharts or reality checks used in the design cycle to help verify that the design and procedures are in place for all components, including the documentation.

Benchmarks may also include timing of various subroutines. For example, benchmarks should be established to ensure that subroutines execute in the expected (and acceptable) amount of time, consume the designed (and expected) amount of memory space, and require the designed amount of operating system intervention and overhead. Execution performance is especially important due to interdependent data communications between the various system components.

Benchmarking also helps to provide logical test points for both internal and third-party testing and certification. For example, if a product is taking a significantly longer (or shorter) amount of time to complete an operation or task than its design dictates, this could indicate a problem that, if corrected early in the process, would be far less costly to solve. Once corrective action has been taken and the solution has been designed and implemented, the benchmark testing and certification process begins once again.

The second step is to test the product's ability to perform the necessary operations in compliance with the benchmark standards and specifications.

Compliance with the Benchmark

Adherence to the benchmark shall be examined in areas of specification compliance, physical properties compliance, and operational compliance.

Specification Compliance

The test team or laboratory reviewing a proposed smart card, terminal, or host system will first examine the technical documentation, including design specifications, operating procedures, and training manuals, and compare the material to the operating requirements and technical specifications that the system must meet.

This examination will include, for example, adherence to ISO 7816 and/or industry-specific specifications such as GSM (telecommunications) or EMV (financial).

Once the test team determines that the technical documentation is complete, the next step is a physical examination of the system.

Physical Properties Compliance

The second test makes sure the system meets specific physical, environmental, and life-cycle use standards. Physical compliance standards require testing for wear and tear (bending and twisting), temperature variation (high and low limits), chemical attacks (water and acid), voltage fluctuations (high and low limits, surges, sags), and clock-speed fluctuations. Destructive tests may include dropping cards and terminals from various heights to see if they still operate after having been dropped. These tests are designed to simulate real-life conditions that physically threaten the system components.

The third area is to examine the operational procedures of the product to ensure that it does what its developers claim.

Operational Compliance

Operational compliance includes testing for software or hardware errors that can be encountered during both normal and abnormal product use. Furthermore, in systems that are using smart cards and terminals for electronic money or other security-intensive uses like medical records, the testing organization may have cards and terminals intrusively tested. Intrusive testing includes invasive physical and electrical probing of the semiconductor embedded on the card, the security module of the terminal, and other sensitive component parts in an attempt to corrupt the integrity of the system.

There have been documented cases of chips and terminal components being physically attacked at the individual circuit level of the semiconductor by electron microscopy and other circuit- and transistor-level techniques. While this type of testing is very expensive, it could identify a poorly designed security feature and eliminate the risk of compromise before the overall system is deployed.

T I P

To reduce the risk of a system compromise at a single weak point, the testing organization should *always* examine related multiple areas using physical and electronic probes and destructive testing.

Multiple-point, intrusive testing is a good idea whether the security feature resides in hardware and/or software. The failure of the test to compromise individual or multiple components of the system increases the confidence level in the whole system. Once it has been established that the chip card can withstand specific attacks and perform required functions, the entire life cycle of the system must be tested.

The life-cycle testing process examines not only the physical aspects of the cards and terminals but the processes each component must go through during its life.

Life-Cycle Testing

This testing exercises all of the functions of the system, beginning with cards before they are personalized. The cards are then personalized, loaded, and used as designed from the first transaction through expiration and retirement. In many of these procedures, a full script may be developed as a part of the qualification.

Often, these procedures are identical to procedures executed internally by the system developers to prove the system prior to its release to a customer. In many instances, an abbreviated or simplified test script consisting of a few critical tests, procedures, and system operations will be used by a customer as part of the system acceptance process. For an electronic purse system, these tests may include loading the card, conducting a transaction, uploading the terminal, clearing and settling the transaction, and retiring the card.

It is important to begin the life-cycle testing process early in the design stages of the smart card application. Testing conducted under real conditions often reveals design faults early enough to correct them quickly and efficiently.

Life-cycle testing and certification must also take into account the coexistence of multiple versions of system software and hardware. The testing process must be able to identify and isolate critical system components and document technical differences, functional enhancements, or new features that exist between versions. These plans can test new components incrementally or by exercising the entire system.

It is important to understand that testing and certification are not simple procedures. A great deal of work and documentation is required to define the parameters and operating specifications of a chip card hardware or software component. The system developer

The Challenges of Multiapplication Cards

The appearance of multiapplication cards, which will become more commonplace, challenges the methodologies used to test single-application systems. The challenge will be to design applications sufficiently isolated from one another so that testing and certification will be necessary only for components added to an already existing and proven design.

To accomplish isolation, you must specify strict parameters for additional incremental applications early in the design phase; the certification methodology must be changed to accommodate the increased complexity. Furthermore, test procedures must be designed and implemented to ensure there is sufficient physical and logical isolation between applications so testing can be accomplished incrementally.

must devise a rigorous plan that will bulletproof an application. A number of organizations or groups are available to assist in developing these plans.

Testing and Certification Organizations

Unlike the telecommunications field, which has well-established (and well-understood) procedures in place, other industries are only now learning and building such expertise. Smart cards require new testing and certification procedures to accompany larger and more complex applications, programs, and uses. The teams that are developing many of these new applications are also developing many of the standards, industry specifications, and procedures. The challenge remains to develop methods to help the customer understand the importance of the certification and testing that will be required for new products.

To meet this need, a variety of industry testing organizations, independent testing laboratories, and certification organizations are expanding their areas of responsibilities to include smart card software and hardware.

Testing Organizations

Telecommunications companies, because of their experience in testing new products, services, and technologies before they come to market, are starting to take an active role in smart cards. Testing organizations such as Bell Laboratories in the United States, GTE Laboratories in Canada, and many national public telephone and telegraph companies (PTTs) have established rigid testing procedures and certification criteria before any network access equipment can be installed in their networks. Their network access equipment now includes smart cards, smart card terminals, and systems.

Independent Testing Laboratories

Testing and qualification procedures are often interactive, using iterative processes established between the developer submitting the cards, terminals, or system for certification and the certifying and testing authorities. This process may include the use of independent testing laboratories. These organizations, such as TNO in the Netherlands, National Physical Laboratories in the United Kingdom, and the National Institute for Standards and Technology (NIST) in the United States, are independent third-party testing organizations that have the engineering experience and expertise to test and certify various types of smart cards, readers, and applications. (See Table 11.1.)

Once a test is successfully completed by one of these organizations, the test report, including the detailed procedures used, can be submitted along with the proposed product during the customer qualification and testing process. This approach can streamline the development process and perhaps cut several weeks off the testing and certification schedule for the customer.

TABLE 11.1   Independent Testing Laboratories

Eclipse
TNO
NPL
Arthur D. Little
BellCore
Dutch PTT

Certification Organizations

Depending on the application, there may also be a governmental or industry oversight authority establishing certification criteria. Financial transaction systems may be subject to stringent criteria established by Visa, MasterCard, Europay, JCB, or American Express. Legal and regulatory criteria may also be mandated by a central bank such as the Federal Reserve and enforced by the Department of the Treasury. Industry organizations such as the German bank consortium, the Central Credit Organization (or ZKA) may establish their own procedures and certification steps to satisfy the requirements of their members. (See Table 11.2.)

Testing and certification of large systems should be approached proactively. For many developers, having products tested and qualified by one industry or internationally recognized testing and certification laboratory may not be enough to prove the viability of the

TABLE 11.2   Certification Organizations

MasterCard
U S West
Visa
Bell Atlantic
American Express
BellCore
SBC

cards, terminals, and systems to another customer. In many countries, various governmental and legal policies may require additional local or regional domestic testing before a system can be certified.

Many international industry associations establish general regional or country-specific guidelines. In most cases, a developer can expect additional testing requirements and procedures to be completed before a system will be available to the market. These additional requirements may not dramatically impact the smart card itself because it is a contained and well-defined component. When testing terminals that use public telephone networks or switched private networks for connectivity of the overall system, though, a developer should expect to encounter an extraordinarily complex and time-consuming process.

It may be wise for the developer to consider methods for reducing risk during this process.

Reducing Risk in the Testing Process

The ways to reduce the risk and complexity of the testing and certification process are as follows:

- ◆ Have well-defined operating parameters and specifications for all product sets, regardless of their size, scope, and complexity.
- ◆ Make these specifications available in an unambiguous, well-documented, and structured form.
- ◆ Have an independent laboratory or third party authenticate the design and certify that a variety of parametric and operational tests have been conducted on the card and/or system.
- ◆ Implement and use internal standards and procedures such as ISO 9000 and 9001 to provide the framework for quality control and quality assurance throughout the design and implementation process.

Most important during testing and certification is how to handle abnormal situations and/or bugs that may appear in one or all system components. There is no single right method to protect against the inevitable software and hardware failures. The developer must design and implement various checks and balances within each of the system components so that exceptional conditions can be detected, reported, and controlled. These internal cross-checks should help provide additional levels of operational compliance and adherence to specifications. In this manner, the integrity of the entire system is maintained and no one component is able to force the system into an inoperable condition.

Testing and certification methodologies usually test for exceptional or unexpected procedure conditions such as illegal dates or unrealistic balances similar to those that may appear if the system's security is breached. The methodologies also specify other checks for possible attacks that would place the system into an illegal working condition. If the system hardware and software are able to trap and highlight these abnormal records or abnormal conditions, the teams conducting the testing and certification can follow the flowcharts, diagrams, and scripts provided by the developer to find the problem. The problem must be fixed by the developer to ensure system integrity.

Procedures for testing and certification are available from many of the card manufacturers, semiconductor manufacturers, and system integrators to provide a base level of component and security testing. These procedures should be submitted to the testing organization, customer, governmental body, association, or consortium that will be certifying the system.

In addition, various component suppliers have prequalified their equipment and these test results should be included in the documentation package delivered to the testing organization.

The Cost of Testing

Finally, be prepared for sticker shock. Testing and certification of a new system is very expensive. In fact, many organizations publish price lists for the various tests required for certification in a given country. Testing procedures should always be included in the system development plan during the early phases. It is also a good management practice to have test procedures in place throughout product design, development, technical walk-throughs, and other checkpoints since they will help lower the cost of the delivered system to the testing and certification authority.

Various procedures and development steps can be used to develop pretests and precertification methodologies during the development and implementation of a card design. The purpose of using this approach is to reduce overall development costs and streamline the certification process. These procedures include the following:

- Use of a testing organization
- Use of proven testing methodologies
- Use of existing hardware and software designs
- Use of pretested software
- Development of components in modular form

Use of Testing Organizations

To hire a testing organization, a developer must first identify which one best fits its needs and will provide the best results. Testing and certification can be so complex that the developer may even subcontract some of its own physical testing work for the proposed product. By involving the testing organization or laboratory from the beginning of the development cycle, the developer can raise the confidence level of the target customer and take advantage of the testing organization's technical insight during critical phases of the development cycle.

Use of Proven Testing Methodologies

Many organizations reuse proven technical methodologies developed to test similar applications, hardware, and software products rather than developing unique, product-specific ones. Since these existing methodologies, processes, and procedures have already proven the commercial viability of other products, it lends credibility to the development process and provides structure to the way the design is implemented, tested, documented, and submitted to an end customer. Similarly, a good development practice is to use existing hardware and software designs as much as possible.

Use of Existing Hardware and Software

These may be designs for a semiconductor memory management program or a communication subroutine that have already been certified in other applications. Using these designs can and should shorten the development cycle time.

You should also consider application code that has already been implemented on the target card or has been tested by others using the same hardware platform. This may be important during the design of product enhancement. An entire certification cycle may not be required for a new functional feature, a software or hardware upgrade, a new mask, or a new security subroutine. Using pretested code may allow for incremental certification. Often, documentation and tests performed in-house during a qualification process by a quality assurance/quality control (QA/QC) department may be sufficient to prove that the system complies with the certification authority's certification process and procedures.

Additionally, consider using as many supplier-developed, pretested software components as possible in the product development cycle.

Use of Pretested Software

As stated before, telecommunication companies have hardware, and possibly software, certified for their applications that could possibly be used to enhance your new system. Although a financial institution may not have much interest in telecommunications hardware, the fact that the hardware has gone through a rigid certification process for each component may be appealing due to cost considerations. For example, the fact that the memory of the component chip or the security module of the system has been tested by one of the certification laboratories often is enough to demonstrate that the hardware has a history and a behavior that is known and predictable. A development team should always request information about the use and certification of components used in other applications so this information can be included in the documentation submitted in the testing and certification cycle.

In fact, all components of the system, including hardware and software, should be checked for any documented testing history.

Use of Modular Components

Low-level routines and simple components should be isolated and verified in modular form to reduce the workload and complexity of bulletproofing the complete application. In this way, the software and hardware primitives, modules, and other components need not be tested individually since they have a long and well-documented history. When the components of a system have been previously certified, only the overall system needs to be put through a process to verify that it is working predictably.

The Role of Engineers in the Testing Process

The final consideration of the testing and certification process is the importance of the engineering staff and its process procedures in the development cycle. Very often, making the development engineers available during the testing or test review steps can help to clarify miscommunications or ambiguous notations and specifications concerning the operation of the overall system. The presence of development engineers is often valuable during third-party certification if hardware and software problems are discovered during the testing of a commercial consumer-type card system.

When looking at testing and certification, you must view the process in the context of the entire design and implementation cycle. Testing and certification must be included from the very beginning of the project. Since approximately 25 percent of the resources of a development effort will be used for testing and certification, the development project manager must be aware of the impact on total system expense if specific attention is not given to this important function. If 25 percent isn't spent early and expeditiously it'll end up costing more later to troubleshoot the system.

CHAPTER TWELVE

IMPLEMENTING AND OPERATING A SMART CARD SYSTEM

A number of complex issues must be addressed before implementing a smart card system. Your organization must decide whether to modify the existing mainframe system to support the smart card application or to build a new infrastructure. Most choose to build a new infrastructure for two reasons:

- Modifications to existing systems can lengthen the development period and thus slow delivery of the system to market.
- Costs will drop, since modifying an existing system can be both expensive and risky.

In many cases, the radical design and architectural differences of smart card systems are enough to justify working outside the existing physical infrastructure. Implementing a smart card system by building a new infrastructure is not a trivial task.

Operating a smart card system is very complex. There is no such thing as a turnkey smart card system that will require no further enhancements. Provisions must be made for incremental upgrades of the cards and terminals in the field, in both the design and the day-to-day operation of the system. Most smart cards are launched with a single application and migrate to a multiple-application environment as soon as business develops. A process must be in place to allow for planned migration as applications mature.

This chapter is divided into two parts. One is devoted to implementation considerations, and the other addresses the operation of the smart card system once it has been commercially launched.

Implementation Considerations

Whether you are replacing a smart card system or expanding an existing system, it is important to consider the following steps in the implementation process:

- Issuance
- Control
- Deployment
- Synchronization
- Initial use
- Expiration and retirement of cards
- Developing a support team
- Marketing your smart card to consumers and merchants

Issuing the Smart Card

During the manufacturing phase, the cards are initialized with application programs, security keys, and other basic information. Personalization, the first step during implementation, is the actual downloading of the *unique* information that distinguishes one card from another. This information may be nothing more than initial balances for an anonymous electronic purse, or it may include detailed information like driver's license data, health records, and bank account or credit card information. The personalization process may be accomplished either centrally or at a remote issuing site.

There are machines and technologies available for the centralized, high-speed, mass issuance of cards. For example, Datacard 9000 is a machine capable of downloading the desired information into the chip, embossing the card, printing the name, account number, and picture of the cardholder, and encoding the magnetic stripe.

At the other end of the spectrum, some cards are issued one at a time. This can be done at a personal computer, a specially configured bank ATM–like device, or a teller machine, using special software to load the desired information into the chip. Slow-speed printers and magnetic-stripe encoders are also available from various manufacturers.

TIP

Issuing cards one at a time may get expensive if you install personalization equipment in a large number of locations. Always consider the cost of the deployment logistics required to support the entire system as you move the cards into the mass market.

The good news is that the lessons learned from issuing billions of nonchip plastic cards (look in your wallet!) will carry over into chip card technology. Adding only one or two processes to existing methods and procedures will put chip card technology within easy grasp of telecommunications providers, financial institutions, and health insurance companies.

The final processing step involves loading cardholder unique data elements onto the card. In some cases, for security and dual-control purposes, this may take two additional steps—only a selected subset of the information is loaded during the personalization process, and the remaining information is loaded later. The approach adopted will vary, of course, depending on the type of application being deployed and whether or not you decide to divide the data. In the types of systems in which loading of the data is split, usually (but not always) the information needed to load the cards is maintained by an on-line central database.

Systems that take advantage of on-line access can accommodate this technique. Since the amount of information that is being loaded initially is limited to a bank balance or credit limit, for example, it can be done in a matter of seconds. The first use of the card by the customer will then download the information necessary to complete the personalization process. This approach provides an alternative to the centralized personalization process.

Regardless of the system you use, it is important during the issuance process to have controls in place.

Implementing Controls

When your system loads data onto the card, careful controls must govern the chip card stock, especially in applications where cards are used as a proxy for value (e.g. money, airline tickets, or electronic traveler's checks). It is important to have careful inventory control not only of the physical chip cards, but also of the other components of the system like terminals and software. When a system is fully implemented with millions of cards and hundreds or thousands of terminals, solid accounting procedures are mandatory from the outset.

Smart card system controls will mimic the control technologies and procedures currently used by the credit card and telephone card industries and include everything from key and lock controls to serial number identification and verification. Physical controls include staging card stocks (so no single location gets too many cards at one time) and spot-checking inventories on a regular basis. On the operational side, it is important to know where each card is located to verify that the cards recorded as issued have actually been issued—especially in large, decentralized implementations.

Card controls must manage how cards are distributed and accounted for throughout the initial issuance process as well as during the operational phases of the system. Further, these controls can check card reliability, especially in multisupplier scenarios where there may be more than one manufacturer, each with different reliability or quality levels. The importance of control to your smart card application extends through the deployment and synchronization stages as well.

Deployment and Synchronization

The implementation process also includes the deployment of the terminals and terminal software. It is important to maintain accurate version controls on all pieces of equipment. Typically, during implementation, even for those systems that start out as pilots and then mature into full-blown rollouts, a reference version that synchronizes cards, terminals, and host software is established by date and/or cycle number. This is important to ensure that the terminals and cards start at some synchronized point, especially for those applications that require security keys.

As discussed in Chapter 4, security key expiration can be either event-based or time-based. Keys, at implementation time, need to start with a derivative of the first or master key loaded in the cards and terminals. It is often a good practice to use a derived key as a security measure to demonstrate that you are running a system that can migrate keys successfully during the operational life of the program. If your system does not use any security features, then this is not an issue and should not be a factor in the implementation.

TIP

Be aware that bottlenecks will occur when deploying cards and terminals. This is mainly because the technology is in its infancy and most deployment processes are not well established.

Once the issuance procedures, controls, planned deployment, and synchronization steps have been finalized, the next step is to monitor the use of the cards during their initial transactions.

Monitoring Usage

In most systems where issuance is remote from a central control point, it is important to have various checkpoints to determine whether the implementation is operating in an expected way. For example, if existing plastic cards are being converted to new chip-based cards, consumer usage must be monitored to quickly identify a technical fault in the card or terminals. Nonusage may also be the result of a lack of product education—that is, the customer does not understand how to use the system. Since spotting and solving problems as early as possible is critical, monitoring system activity during the first few weeks of deployment is essential. At the other end of the spectrum, expiration must be considered.

Expiration

Another important checkpoint during implementation is the expiration or retirement of cards and/or terminals. This should be tested either through a pilot or in an intermediate cycle early in the deployment program. The test should be designed to verify that when an event and/or a usage-based expiration occurs, the sampling techniques in place validate all of the life-cycle issues during the implementation process. Before implementation, however, a support team must be in place to ensure your system's success.

Support Team

It is important to have extra team members available to monitor help desks and troubleshooting locations to ensure a smooth implementation of the operating environment. The assigned staff should be well trained, and complete documentation should be in place to help them solve operating problems on a day-to-day basis.

In the migration of an application from a nonchip-based to a chip-based system, the easiest way to implement the change is to do everything that was done with the older technology, whether it was based on a magnetic stripe or not. The new system features and abilities are simply added on top of the base functionality. For the foreseeable future, we will live in an environment composed of many different types of card technologies. Thus it will be important that customers understand how to use the newer technology and that the terminals accept multiple media cards. The design of the new services and products should reflect the existing ways customers interact with current products, such as the method of inserting the card into the terminal and the way in which messages or prompts are presented on the screen. The new customer interaction process must remain easy to follow and as intuitive as possible. Once the application has been converted, the difficult work of operating it over the long term begins. For more detailed information on implementation team members, See Appendix E.

Marketing Your Smart Card

After you've thoroughly tested your system and confirmed that it's secure, you must convince both merchants and consumers to accept and use it. Your marketing staff must concentrate its efforts on marketing to consumers, and your strategy should focus on the following:

♦ Educating the consumer
♦ Assuaging consumer fears
♦ Identifying with an established brand name

Customers will not use a product they're unsure of, and most people still feel insecure about handling their money electronically. To combat this, you must consistently emphasize that your product is safe and secure. The best way to accomplish this is to explain your complex system as simply as possible, so that even the most technically ignorant consumer understands fully what your system does and how it works. Although the growing popularity of shopping on the Internet has helped to ease the transition to a cashless society, helping the consumer understand your product will be a perpetual problem as you roll out or upgrade your smart card system. As you're educating your customers about your system, though, you must be careful not to push your technology-is-good message too far (see the Utah driver's license case study later in this chapter).

While you're explaining how technology will make your consumers' lives better, you must be prepared for an antitechnology backlash. People still harbor strong fears of living in an automated society, and, as you'll see in the ensuing case studies, Luddites can influence the acceptance (or lack thereof!) of a smart card system. One of the best ways to overcome such anxiety is to identify your system with a brand name.

As the current VisaCheck Card advertising campaign illustrates, people will use a product if it comes from a company they trust. In successful smart card rollouts, smart card companies can market their products to banks with strong brand-name recognition, and, once aligned with them, use their resources and reputations to overcome consumer—and merchant—fears.

Marketing to Merchants

You must also create a plan for convincing businesses to use your product. In doing so, we recommend that you follow these guidelines:

1. *Never assume immediate merchant acceptance.* No matter how much the public loves your product, unless businesses are willing to accept it, it'll be about as useful as an eight-track cassette player. (We listed this guideline first, despite the fact that the other guidelines are arguably more important, because it's the easiest to overlook.) The VisaCash card used at the 1996 Olympics was felled partly because card designers were unable to convince merchants outside of the immediate games area to accept the card.

2. *Use existing hardware.* All smart cards will still have to work interchangeably—there's only so much room for card readers at any merchant's cash registers! Given

today's climate, though, if your smart card works with equipment that's already standard for the vast majority of businesses, the odds of a merchant agreeing to accept your card are exponentially better.

3. *Identify with a brand name.* As with consumers, established brand names have a better chance of being accepted by businesses because the relationships are already in place.

Case Studies

The following four case studies (an attempt to use a smart card as a driver's license in Utah, the VisaCash program at the 1996 Olympics, the Mondex electronic purse system, and the Banksys Proton system) show how the success of your project hinges on both marketing and technology.

Utah's Driver's License

As we discussed earlier, people fear change, especially when it involves something they do not understand. Utah Governor Mike Leavitt discovered this the hard way when he attempted to implement a smart card–based driver's license.

The Utah driver's license project failed because it didn't explain the system well and it failed to assuage consumer fears. When these two principles are not addressed properly, people tend to make wild assumptions about your technology.

Utah wanted a multiple-application card that included the driver's license as a base application and loyalty and financial applications as add-ons. The design team used an innovative approach to help achieve these goals, creating an 8K EEPROM smart card with 2K reserved for the state and 6K for sale or rent through multiple branding.

This project died because a coalition of citizens' activist groups, including the ACLU, opposed the initiative because they feared the state would build a database of people and/or otherwise use the information in violation of the Constitution. Other groups, such as conservative religious groups, feared the technology as the work of the devil. It was even reported that the chip could be used to track Utah residents by helicopter. One particularly adamant group wrote numerous letters against the plan, identifying themselves as the "Citizens Against Smart Cards."

Because the marketing efforts failed to educate and reassure the population, these voices of hysteria drowned out more moderate, rational discussion of the technology. This is not to suggest that fringe elements will agree with you once you fully explain your product or that it is possible to convince everyone that their concerns have been addressed. However, a marketing plan focusing on exactly what the card could and could not do would have gone a long way toward shaping popular opinion in Utah.

The Atlanta Olympics VisaCash Trial

The 1996 Atlanta Olympics was the stage for the first stored-value smart card program in the United States. Visa International's (the promoter of the Atlanta project) strategy was to use Visa's brand recognition to market to both consumers and merchants. Visa recruited popular banks like NationsBank, First Union, and Wachovia Bank to help implement its system. NationsBank used its brand recognition to convince the major Olympic venues—Olympic Stadium, Georgia Dome, Georgia World Congress Center, and Alexander Memorial Coliseum—to accept the card.

The card itself was a single-application electronic purse, designed to be discarded after all the value was decremented (ironically, these cards are now considered valuable by card collectors).

The pilot project ran into technological snags early on that diverted its attention from marketing. Visa had licensed its hardware from a Danish firm, and their programmers quickly spotted a flaw in one of the chip suppliers' products. After solving this problem, Visa ran into trouble with its readers. Eventually, Visa and the banks debugged their software and hardware and were ready to launch their system.

At the Olympics, when consumers tried to use the card, they found Visa had failed to mention that the VisaCash card was *different* from a standard Visa credit card, and they also had failed to alert consumers that merchants who accepted Visa credit cards might not accept VisaCash cards and vice versa. After while consumers became very frustrated because they could not use their cards or had to fumble around their wallets or purses to get the right Visa card. In addition, signs showing which cards were accepted where were very similar, adding to the confusion, especially for foreign guests. Finally, outside of the main Olympic venues and contiguous areas, Visa hadn't convinced merchants to accept the card, and the cards were worthless. This became a real issue when people ventured outside the general Olympic area.

Despite the fact that VisaCash assuaged consumer fears by marketing its system as a convenience device, it failed to properly educate consumers or to convince merchants to accept it. Now, though, the VisaCash rollout serves as a working model for virtually all future smart card rollouts in the United States.

Mondex

Mondex is an electronic purse system designed in Britain by NatWest Bank to imitate cash. Unlike other electronic purses, Mondex was unauditable—its transactions could not be traced. Mondex used this aspect of the card as its major selling point in order to quell customer fear and spur its acceptance, since there was no potential for data mining or other invasions of privacy. Mondex even allowed wallet-to-wallet and card-to-card transfers. Consumers could exchange money via the card just as they could exchange cash.

As good as it sounded, though, the Mondex technology was incredibly risky. It was especially vulnerable to money laundering, injections of black-market cash, and tampering. Mondex overcame this with a phenomenal marketing campaign that assuaged fears and convinced large banks to issue cards in spite of the technology concerns.

MasterCard, Wells Fargo, Royal Bank of Canada, CIBC, NatWest, and HSBC all adopted Mondex, and MasterCard eventually purchased 50 percent of the company. Interestingly enough, Mondex convinced banks to invest in its questionable technology by marketing its own product as *the* industry standard for electronic purses. Statistical evidence suggests that Mondex was not as experienced as marketers led the investors to believe.

MasterCard plans to use the Mondex platform in New York, but must change its internal architecture to make it auditable in order to comply with the Federal Reserve. The new American Mondex will be fundamentally different from the smart card systems we've just described, but this case illustrates that careful adherence to the marketing guidelines allowed even an intrinsically faulty system to become a commercial success.

Banksys Proton

The Banksys Proton system was designed in 1993 to serve as the national electronic purse for Belgium. After two trials, the system rolled out nationwide, and additional implementations followed in Sweden and the Netherlands. In November 1996, American Express purchased the rights to use the Proton product on a nonexclusive worldwide basis.

The system, a complete end-to-end system, was packaged extremely well. Of the major electronic purses commercially available, Proton is one of the most cost-effective security cards and is attractive to many banks because it has good back-end interfaces to the banking host and processing systems.

In addition to its technological perfection, Proton is the most successful electronic purse that currently exists. Rather than create a brand name of its own, Proton has aligned itself with companies that have strong name recognition within their respective countries, and with American Express, which has global recognition. Like Mondex, Proton emphasizes that its applications make life better for consumers, and this approach quells consumer uneasiness. For example, it highlights the fact that low-cash-value transactions that frustrate customers, like vending machines or parking meters, become convenient and easy to manage using Proton. Proton has also approached vendors early and often, using its local brand connections to advantage.

Operational Considerations

To determine operational considerations of smart card applications, begin by looking at the implementation steps; the initial implementation steps over time become the operational steps. For example, the processes of upgrading cards and terminals in the field or upgrading host software can be viewed as continuous mini-implementations that enhance the capabilities of the card or terminal to meet the changing needs of cardholders.

Additional important operational checkpoints must be put in place to verify that the smart card application is working as specified and that system integrity remains intact. Operational considerations should include card replacement, system integrity, changing software in the field, expiration management, management reports, and problems that may be encountered in changing operating environments.

Card Replacement

The first operational consideration is the process of card replacement. Card replacement will be necessary when a customer's card is lost, stolen, destroyed, a victim of electronics or software failure, or any number of other accidents. Whatever the cause, it is important to verify the reason for the reissuance if at all possible. Verification is simple if a cardholder reports a card lost or stolen. But if the card is returned to the system operator or is otherwise determined to be inoperative, it must be evaluated to determine whether the technology is at fault.

Reconstructing the contents of a card when reissued is also an important operational consideration. In many of the systems described in this book, the card itself is only one of the locations where the data is stored and managed. Redundant data repositories provide the system operator with the ability to re-create the distributed database on the card from a source managed by a computer higher in the system's hierarchical infrastructure, such as a host computer.

T I P

Returned cards must be tracked over time to ensure that a pattern of hardware, software, or manufacturing defects does not exist, or if a pattern emerges, the cause is identified and removed from the operating environment.

In this instance, the system operator can reconstruct the card once all the transactions are forwarded to the central database and the value is recalculated. The next task is to create checkpoints that will verify the integrity of the system and of these cards.

System Integrity

The system operator must continually check to confirm that there are not more cards being used than issued. The operator should also check that individual card balances (if applicable) do not exceed the amount for which they were issued. From an audit perspective, the comparison of a card-based variable against a value stored on the host computer verifies not only the balance or value information, but also some of the keys. These types of operational checkpoints help ensure the system is operating according to its specification.

Additional checkpoints in the audit process include the random sampling and a recall policy requiring that cards be returned by customers who relocate or discontinue service.

T I P

Destructive testing is a common practice in the card industry; it is used to better understand the operational characteristics of products over time. Destructive tests of a random sample of returned cards may range from bending and twisting the cards to using them to remove ice from car windshields. The results of such tests can help gauge the card manufacturer's quality.

Another area of operational concern is changing the software or security programs in the field, since this activity can be a very complex process.

Changing Software in the Field

Changing software in the field should not be attempted until you have carefully planned the process and determined the controls that will be necessary. When software changes are required in existing cards, the physical exchange is accomplished by issuing new cards rather than reloading program data into an existing card. The reason for this is that existing first-generation applications tend to use up all the available program and memory space in a card.

However, for many future applications, the reservation of additional data space may allow an extra field of data or an extra security key to be loaded on the fly while cards are in commercial circulation. In these cases, a methodology to load the variable data or key must be designed and implemented to allow the introduction of incremental units into the system.

As described in Chapter 10, most card applications in use today, with few exceptions, have an on-line synchronization step. The synchronization step is very important not only to verify the integrity of the system during a value reload or redemption of loyalty points, for example, but also to load new keys, security algorithms or procedures, or, in this case, new applications. When the loading process is successfully completed on-line, the verification can be immediately reflected in the host computer. (See Figure 12.1.)

From a monitoring and control perspective, it is important to know which cards are in use or not in use and which have received upgrades. This information must be known when planning to update terminal or CAD software to a new version level or to stop supporting an earlier level.

A card that has been idle for more than six or eight months, for example, may be forced to execute an on-line transaction before it can be used again at a terminal. The on-line session may synchronize the keys, load various transactional information, or load new software to allow the card to be used with terminal software that

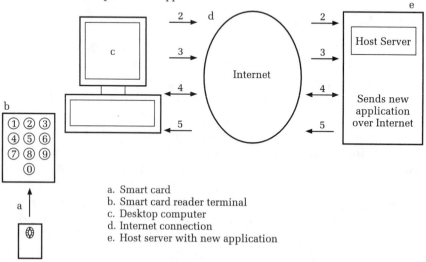

1. Card reader authenticates card (shut off from Internet connection for security)
2 Reconnect to Internet
3. User requests application update from host via Internet
4. Host server authenticates user
5. Host computer sends application

a. Smart card
b. Smart card reader terminal
c. Desktop computer
d. Internet connection
e. Host server with new application

Figure 12.1 Card and terminal revision and generation acceptance.

had been upgraded to a higher revision and operating level. This upgrade technique, however, could take years in some applications, or days, weeks, or hours in others—for example, those being used by pay television and satellite reception systems, where synchronization of keys and key management may occur every few seconds. Revision can occur many times during the lifetime of a card and must be considered throughout the entire life cycle of a commercial system—right up to the moment that the last issued card has expired.

Expiration Management

Card expiration will be based on one of the following:

◆ Time

◆ Exhausted value

◆ An event

When a card expires, several things may happen. A locking function may be programmed into the card to make it inoperative for a specific application. Another method is to program the card to erase keys or key information. Other program features may place the card in a read-only, nonupdate mode or simply cause it to self-destruct. The approach used will depend on specific application and expiration requirements. As discussed before, the method used should be verified by testing a small population of cards during the initial issuance process to verify that those steps are in place.

Expiration management becomes a complex issue when considering a multiapplication card. For example, in a card for a telephone application, monetary value to make telephone calls may be contained in one section of the card, while another section contains loyalty points. The telephone section may expire after a certain amount of time, while the loyalty points may remain on the card indefinitely, or until another expiration date occurs. At issue is when to expire the card and render it nonfunctional. The system design should include the provision to renew cards based on their ability to take on additional information. The card expiration date could be reset or the loyalty point information could be transferred from the expired card to a new one issued to the cardholder.

As discussed earlier, the issuance step, which may require on-line synchronization, becomes especially important and takes on a very critical role in the implementation of systems that require multiapplication technologies. In this case, the synchronization of all the different applications is accomplished and the card is updated with the latest and newest information. All the data from the expired card migrates via the central computer onto the new card. In multi-issuer or multihost systems, this may require that online changes not be limited to a single issuer. (See Table 12.1.)

TABLE 12.1   Multiapplication/Multi-issuer Requirements

Multiapplication	Multi-issuer
Large memory	Database management
Upgradability	Secure application design
Application independence	Internet access capability and upgradability
Coprocessor for cryptographic security	Security management for lost, expired, or stolen cards
Management of multiple expirations	Synchronization of multiple applications
Conformance to ISO standards	Conformance to ISO standards
Interoperability	Interoperability

In this case, a central coordinating function must be established to define the relationship between the card and its primary and secondary applications. When a consumer loses a card, for example, he or she will look to a primary issuer for a replacement. In the example of a government services card—perhaps the electronic driver's license of the future—the primary issuer may be the Department of Motor Vehicles. This issuer would be responsible for notifying other departments with applications on the card—for example, the Election Commission for voter registration, the Department of Health for immunization records, and Medicare for health care information. In this example, the Department of Health would not have access to the motor vehicle database and vice versa. The primary issuer's responsibility would be to notify the owners of other applications on the card that an update of the information on the reissued card was required.

When multiapplications belong to competitive companies, notification may be accomplished through secured digital certificates managed by a neutral third party. The third party holds the certificates to ensure that all the information required to resynchronize the reissued card is loaded securely and confidentially. If the activity is accomplished by the U.S. Post Office, the effect would be similar to sending electronic "registered" letters to the proper recipients. There will be times, however, when expiration is not a controlled process but happens unexpectedly due to a catastrophe.

Disaster Recovery

As commercial systems improve in capability, function, features, and complexity in response to competitive pressures or regulatory changes, there will be application upgrades over time. It will be important to have a process in place, including a contingency process, to handle the reissuance of some or all of the cards in a system.

These plans will have to handle various situations, from loss or catastrophic failure of one or more of the components of the system to the threat of attack or security breach in the cards or terminals. The backup measures that must be in place are similar to the disaster recovery plans of central computer sites. The advantage of using distributed computing (in this case, down to the consumer level where each card is a part of a overall system) is that the system integrity has a better chance of surviving any one single attack. In the event that the system uses asymmetric security, even the security algorithms are virtually protected from attack. A system audit maintained throughout this process will monitor the operational information on the card and terminals in the event a switch needs to be made to a new security key or a reissued card.

In addition to disaster recovery, management reports are an important deterrent to catastrophe.

Management Reports

There are few unique or new reports that do not already exist to control card-based systems, with the following exceptions:

◆ Tracking reports to monitor failures of various cards
◆ Lost-card reports to monitor value and to restore value to a cardholder

- ◆ Lost-card verification reports to track, for example, card-holders who frequently lose their cards
- ◆ Systemwide product-fault reports to monitor and isolate problems by card, terminal, or manufacturer

Other reports generally included in card-based systems are over-all reports, which include card utilization and usage, security alerts, application-specific information, and available memory.

Card-Usage Reports

The way in which cards are used has an impact on the number of security keys, versions of the hardware and software in the cards and terminals, and the concurrent types of software that must be monitored. This information is important to ensure that all cards will get a response from a terminal when they are presented for use. Infrequently used cards should be able to communicate—even if the message directs the cardholder to take the card to an issuing office to have its keys refreshed or additional applications loaded.

Security Alert Reports

The security alert or exceptions report lists keys being attacked and security breaches detected by cards or terminals. Terminals and cards should always record and report a faulty authentication. Faulty authentication may indicate that an illegal terminal has been introduced into the system by a potential criminal, or it could mean something as benign as a failed terminal. It may also indicate that a card is being simulated. The transaction event should be recorded by both the card and the terminal and reported to the system operator. These kinds of transactions will have their own sequence numbers, just as a monetary transaction does, and can be traced as an informational transaction that will go onto an exception report for research and resolution.

Application-Specific Reports

Application-specific reports track such things as loyalty points earned, loyalty points redeemed, and airline tickets issued. These reports also track the number of certificates or keys used and/or migrated. There should always be a buffer inventory or some safety stock of nonissued or nonutilized keys or key methodologies, both in the security modules of the terminals and in the security sections of the cards. In the instance of an attack, event, or time-based trigger, the terminal or card would initiate a transaction to activate new keys or key registers. This capability is especially important for applications like secure electronic commerce, where the cards are acting as electronic tokens or signature and validation devices. In the case of marketing-only databases, event recording may be of less interest or value.

Available Memory Reports

Some applications may be transaction- or data storage–dependent. Many applications, as described in previous chapters, will use fixed-size files and data structures like cyclical files to keep the most recent transactions in the card memory. However, others may be life-event-recording applications, such as student transactions in a university environment, where the amount of data in the card may at a certain point exceed the existing resources. Reports have to be developed to measure and track the resources available in cards and to ensure that the card resources are being used as the application specifies. If it is found that resources are being used incorrectly, cards with increased memory may have to be issued (at additional cost) or possibly redesigned to expire certain data in order to free up memory.

TIP

Reports compensate for a project manager's inability to consider every conceivable scenario in the initial design. The operational measurements used to calculate the effectiveness of the system must be flexible.

As can be seen through an examination of various reports, the advantage of using smart cards is the ability to have more than one way of verifying the integrity of the system. The ability to recall cards, check on terminals, and use management reports to ensure that the host system is in synchronization with all the other pieces of the environment affords greater control and integrity of the system.

T I P

Controls are not necessarily able to validate the content of the transactions or the information stored within the card. Control of keys and private **information,** especially in multiple applications and multi-issuer environments, must be preserved through use of strong security techniques such as certificate-based encryption. Such techniques help ensure the cardholder's privacy throughout the life of the smart card application.

However, as smart card systems proliferate, reports will not be enough to monitor activity across state lines or national borders.

Operating in Different Kinds of Environments

Most of the systems in place today tend to be regional or national in scope—for example, a closed system that operates in a university or an electronic purse system that operates in a town or even in an entire country. As we move to more sophisticated systems and applications (such as electronic commerce and the international applications contemplated by Visa, MasterCard, and American Express), we're going to find increasing demand for ways to provide for uniform data compliance across different country-specific legal and regulatory requirements. In this expanded operating environment, the card may be only a placeholder of certain important elements of information and not the sole processing, storage, or key component. The card would allow, for example, a transaction to be conducted in one country, processed in another, and fulfilled in a third. Thus the legal ramifications of the transaction may not be as easily definable as in existing systems.

While this is not meant to create a loophole scenario, it is important to have an appreciation for how commerce will be conducted in a community of electronic environments. When advanced card systems, such as those using chips, expand beyond a single border (which will also be true for global electronic commerce), the system is no longer constrained by the physical, legal, and regional boundaries that have been the rule in the past. The PC and the Internet have tremendous implications for the legal operation of the system. Applications that cross borders may have multiple scenarios of security, validation, transaction recording, and receipt requirements designed into the cards, depending on where they are being used. Differing transaction or value-exchange processes may be enabled in real time to comply with the local legal and regulatory environment.

It is important, therefore, to keep an eye toward the future, carefully considering the operational logistics and requirements of having multiple options available in the cards. For example, PIN numbers, required in many European countries may be required in the United States also in an electronic commerce transaction conducted with a non-U.S.-based merchant. In this circumstance, the PIN would be required or the card could substitute a proxy (stored PIN) that would be accepted in the foreign country. These transactions will probably have to be cleared through a third party that has a trusted relationship between the two countries.

This situation creates the potential for an entirely new industry arising to handle electronic commerce through a chip card. Keep this in mind as multiple-use cards become increasingly popular for everything from value exchange, transportation, telephone and Internet communications, and public services, to things such as electronic airline tickets, loyalty points, and other travel-related information.

CHAPTER THIRTEEN

THE FUTURE OF THE INDUSTRY

Credit, debit, ATM, loyalty, and membership cards have changed our lives and the way we conduct business. We use them to obtain cash, track frequent flyer miles, and gain access to secure areas. We anticipate the formation of entirely new business organizations based on smart cards to be launched in the next millennium. In this chapter we describe several possible future scenarios that smart cards may enable.

The smart card industry is quickly maturing, moving from an analog physical card environment to an electronic one. As a consequence, an entirely new type of commercial landscape is being created. This environment will be far more secure than today's magnetic-stripe- and paper-based transactions. Future trends that we expect to see include the following:

- Electronic commerce
- Improved technologies
- New time constraints
- New skill sets
- Easy-to-use applications
- A merging of disciplines

 ◆ An increased role for biometrics
 ◆ A growing array of appliances
 ◆ More sophisticated security
 ◆ A greater diversity in the development and innovation of smart card applications

Electronic Commerce

We predict there will be new kinds of information-access scenarios available in which the consumer and the merchant will not have to be physically colocated to conduct commerce. This means we will buy things wherever and whenever we want to, without the requirement of being physically present at a car dealership, grocery store, or restaurant. The consumer, not the merchant, will assume electronic control in many future transactions. Some scenarios predict that as many as 85 percent of the new terminals and card-accepting locations under development will be user-driven.

Our TV sets (with over a billion manufactured) and telephones (over a billion phone lines) of the future will have the ability to accept smart cards to enable electronic commerce. There are already millions of digital cellular telephones that accept these cards today. New personal reader devices, such as the Visa viewer, can display your electronic cash balance and your last 10 or so electronic cash transactions. Other devices will evolve as more sophisticated personal commerce and information-access devices become available.

All these devices will be our "PCs" of the future. They will be more than just entertainment devices (like TV sets), points of communication (like telephones), or places to work (like personal computers.) These "smart" devices will become points of interaction and electronic transaction in tomorrow's economy. A securable device like a smart card will be important in enabling this vision to become reality. We are at the beginning of a new awareness and a new environment.

Improved Technologies

In setting expectations for future opportunities, you should expect the unexpected (that is, "future-proof" yourself). The smart card industry today is fairly well defined by existing levels of computer technology. We expect to see smart card technology define a path and evolution similar to that of the PC. For example, today's smart cards and those being designed for use through the end of the decade operate at half-duplex. They can either transmit data or receive data, but they cannot do both simultaneously. Tomorrow's smart cards will become full-duplex cards with the capability of transmitting and receiving data simultaneously.

These future cards will also run at higher speeds, be able to perform more secure functions at lower voltage, and have greater storage capabilities—much like the personal computer of today. More and more of today's features and functions will grow and mature into far more sophisticated systems. Technology and time constraints will be linked in the future.

Time Constraints

The time constraints that are defined for a product, even to the point of setting an expiration date, may have nothing to do with the realistic lifetime of the card (i.e., the number of cycles of the electronics) or the degradation of the plastic while it is in a wallet or purse. The constraints will be defined by the next generation of technology, or the need to upgrade to more sophisticated applications. A good analogy is today's PC environment moving to faster and more complex microprocessors with more processing, memory, and storage capacity to support the multimedia and Internet TV applications emerging today. It is very likely that the smart card and PC technologies will converge in the near future; the card may become an electronic key to unlocking PC-type applications.

Building smart card applications will not require new technical skills or exotic university degrees.

Skill Sets

Future skill sets will be based on an incremental evolution of existing skill sets and disciplines, including database design, distributed computing, networking, security, cryptography, and so forth. The team required for a system implementation must blend these technical skills with an understanding of smart card capabilities and operational considerations. At the mask design and implementation level, a programmer or analyst who has experience with real-time programming can pick up smart card skills in a matter of months via incremental training by existing semiconductor and card operating system manufacturers.

Programming skills for real-time terminals or consumer electronics will be a requirement in support of smart card applications.

Even in this area, most of the new experience and knowledge required will be in the security area due to the fact that today most consumer appliances do not have very sophisticated security algorithms or features.

Most appliances today are read-only, even if they use a magnetic stripe or chip card. New appliances are being designed in a read/write environment where data may be dynamically updated depending on the application and how the information will be used by the appliance.

From an overall project management perspective, the skill sets required will be those that are taught today in classical systems design and systems implementation courses. The most important thing to understand is that in a store-and-forward system, data is aggregated at various levels and at various times in the transactional life cycle. Additional understanding is required in the operations area, because operating these card environments requires a technical level of project management. The project manager needs to understand, from the engineering perspective, all of the potential failure modes and faults of cards, terminals, and other equipment and have the skills and resources to track down security faults or breaches. Mastering operational knowledge and skills will probably require a 6- to 12-month learning curve.

Finally, making these applications easy for the consumer to use and understand will require a significant amount of thought in the design and implementation of the system.

Easy-to-Use Applications

Smart card applications go far beyond the requirements imposed with the migration from nonmagnetic-stripe applications to magnetic-stripe applications. In this case, the level of training (to introduce new skills to the existing merchant workforce and individual consumers) needs to be carefully analyzed and budgeted in the overall cost of the system. Increasing the abilities of the frontline staff to operate smart card systems is an important element that cannot be ignored in the implementation phase.

Merging Disciplines

Electronic commerce will require merging disciplines in several areas, the most important of which is security, as these new transactional systems will increasingly rely on more sophisticated devices. These devices will have various formats and form factors, ranging from contact and contactless chip cards to PCMCIA cards and other personal electronic tokens such as advanced automobile keys, keyless entry systems, or secure identification technologies.

The role of biometrics (security uniquely tied to each individual) for identification will increase dramatically.

Biometrics

In the future, biometrics (e.g., fingerprints, eyeprints, voiceprints, hand geometry, or some other unique physical characteristic), coupled with other information known only by the specific individual (like a PIN) will be stored on the smart card. The use of biometrics will allow electronic signatures and transaction authorization. Scenarios of the future (and in fact, in use today) may include such applications as border control, with the United States Immigration and Naturalization Service using an advanced entry system called INSPASS.

INSPASS is a rapid-entry system that uses a normal plastic card with optical character recognition (OCR), bar codes, and a hand geometry reader. Future systems will use a smart card to securely store a fingerprint or handprint such that the card can be used at any border or control point, even when the reader is not connected to a central computer.

Today's systems use on-line verification of cardholder information; tomorrow's systems may or may not be on-line. In securing the information inside the chip, specific biometric data can consume a significant portion of the memory. Not only is it important to have this information securely stored in the device, but its ability to be altered, changed, or modified must be very strongly protected. Protecting that information is a part of the role of the smart card chip once the data is entrusted to the card.

Biometric applications also require a way to securely transmit information into and out of the card. Possibilities for future systems include a proposal that biometric information, such as a fingerprint or a handprint, be transmitted to the card from the reader for comparison with some computational algorithms stored inside the chip. In either method, whether an encrypted version of the information is sent out for comparison or the comparison is done by the chip card or both, advanced authentication techniques, including biometric information, are on the horizon. (See Table 13.1.)

Future applications include the idea that these cards will become an integral part of an increasing array of appliances.

The Growing Array of Appliances

It would not be totally unrealistic for appliances (ranging from TV sets to refrigerators) to be equipped to read smart cards. In the case

TABLE 13.1   Biometric References

Biometric Consortium	www.vitro.bloomington.in.us: 8080/~BC/
The Association of Biometrics	www.npl.co.uk/~dsg/afb.html
Recognition Systems: articles	www.recogsys.com/articles
The Biometric Digest	phone: (314) 851 0924
Biometric Technology Today	phone: 44 458 27 44 44

of a refrigerator in a public area, a card could be used to unlock the unit and record who accessed the device and what was removed. An industrial application could be the similar accessing of commercial chemical-storage areas.

The greatest growth in card-equipped appliances, however, will be on college and university campuses. Cards will provide access to and stored-value exchange for vending machines, TV sets, washers, and dryers. As a result of these applications, we will see increasing numbers of personal readers and personal electronic commerce devices. (See Figure 13.1.)

The number of these devices may some day equal the number of cardholders (there will be more than one card and chip per person, but probably only one card reader per person).

With this immense growth in readers and points of access in the market, we will see many other systems launched to take advantage of this electronic future. As we described earlier, only 15 percent of the access points of the future will be established as merchant-type points of interaction, and 85 percent of the future access points will be consumer, self-initiated devices. The very numbers of unattended devices raises the issue of security to new heights.

Security

The new security threats caused by the evolution of technology have been anticipated by existing applications developers and card system operators, who are using colleges and universities as testing

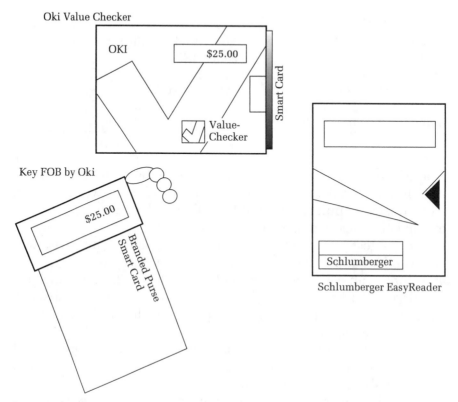

Figure 13.1 Examples of personal smart card readers.

labs to study the nature of these attacks and how to thwart them. Often, the main threat of attack to defeat or interrupt the operation of the system is not from the causal hacker. The greatest threat is from high-tech organized crime (organized crime has tried repeatedly to penetrate monetary and other high-yield systems). Therefore, card and terminal design must have more sophisticated logic and security capabilities, such as certificate authentication and mutual authentication procedures.

Even the personal reader devices would have smart card–complementary logic in order to provide part or all of the authentica-

Elliptic Curve Cryptosystems (ECC)

Fraudsters have improved their ability to crack smart card systems by using more powerful computers. This situation created the requirement for longer key sizes for most current public key systems. However, longer key sizes make these public key systems even slower and more cumbersome.

Elliptic curve cryptosystems (ECC) are analogs of public key cryptosystems such as RSA in which multiplication is replaced by elliptic curve addition. ECC is being strongly considered by standards developers as an alternative to established standard cryptosystems such as RSA. Some are calling elliptic curve cryptosystems the next generation of public key cryptography because they provide greater strength, higher speed, and smaller keys than established systems.

One of the main advantages of the ECC is that the arithmetic is easier to implement, both in hardware and software. ECC is particularly beneficial for applications in which computational power and integrated circuit space are limited, high speed is required, intensive use of signing, verifying, or authenticating is required, and signed messages are required to be stored or transmitted.

Fast, strong, ECC signatures can be performed on standard chip cards without the need for a coprocessor. ECC also minimizes code, key, and certificate storage space to allow space for the application. ECC also provides the lowest-cost deployment of chip cards in the shortest time frame.

ECC has been implemented by various groups around the world:

- Matsushita (Japan)
- Certicom Corp. (Canada)
- NeXT Computer (United States)
- Siemens (Germany)
- Thompson (France)
- University of Waterloo (Canada)

tion between the card and the card-reading device. This implies that the number and complexity of readers will, at a minimum, have the same capabilities as the complex cards in the market. From a system design and architectural perspective the readers will have to be treated as discrete units in the commercial operating environment and may have their own unique cryptographic and addressing methodologies.

This approach is not all that dissimilar to the Internet system architecture of today. However, we must take care to allow sufficient ways to assign unique identification numbers, keys, or certificates of the components in the overall system so that we do not run out of identification numbers or addresses, as evidenced in today's Internet and other large, connected, network environments.

Future Applications

Smart cards will quietly revolutionize most sectors of the economy. Many cards will find a core application, with additional applications being built on to the primary delivery system, such as an electronic driver's license with state and federal welfare programs launched on top of it. A comprehensive bank card may have applications ranging from merchant loyalty programs and electronic purses to debit, credit, and possibly consumer finance and lending. We also expect to see ticketless travel and frequent flyer/driver/sleeper programs on multiple-branded cards. Only our imaginations limit innovative applications of the future.

At a certain point, though, the competition and demand will dictate the applications that are developed and implemented. From the technical side, it is a question of making the appropriate trade-offs between the various technologies and being flexible enough to accommodate the future systems and technologies that will evolve on-line (or off-line).

When considering the trade-offs, we must also take note of the increased risk of putting too much information onto a single card. For example, over the next few years, JavaSoft plans to incorporate other capabilities into the Java Card API as applications become more sophisticated and the feature sets of smart cards expand including the following:

- Unicode character set
- 32- and 64-bit integers
- Float and double-data types
- Unidimensional arrays of unsupported data types
- Multidimensional arrays
- Arrays of objects
- Exceptions
- Threads

The risk here is not a technological one, but rather a business and logistical one. The competitive information contained on such a card from members of the airline industry or the retail merchant sector will make a lost, damaged, or expired card very expensive to reissue. The presence of more than one application on a card will make it difficult for the individual cardholder to reconstruct the information on a lost or damaged card or to update all the information on a reissued card. Furthermore, the problems of accessing the enormous numbers of cards is daunting as they proliferate to meet the requirements for products and services of individual cardholders.

Staying Ahead of the Game

A final area to discuss is finding a process or methodology to maintain a lead on the developments and innovations in the smart card industry. Implementing an application and making the commitment to go forward with chip cards is really buying into an entirely new business process, which must involve the entire enterprise from the beginning. Once the initial education process is completed and an application is launched, a methodology must be in place to keep abreast of the latest industry developments in products, innovations, applications, toolkits, learning, dynamics, and the legal and regulatory issues as cards are used for government applications, consumer electronic finance, and the like. Keeping on top of the applications is going to be a team requirement, and all employees, from top management to line managers who operate the system, need to be continuously aware of developments in the technology and in the system.

Lead times from innovation to realization continue to shrink. Increased knowledge (where knowledge is power) will be the key element to sustaining a competitive edge in these emerging applications. When you issue consumer electronic devices such as smart cards to replace well-defined and historic instruments such as coins and cash, the system has to operate flawlessly and reliably from the onset. Having all parties in an organization understand what needs to be done now and what will need to be done as these systems evolve is going to be an increasingly important activity.

GLOSSARY

asynchronous A data transmission scheme that handles data on a character-by-character basis without clock synchronization. The character code normally includes a start bit that indicates the beginning of a data character, 5 to 8 data bits, an optional parity bit and 1, 1½, or 2 stop bits.

card acceptance device (CAD) Device used to communicate with the ICC during a transaction. It may also provide power and timing to the ICC.

certification Endorsement of information by a trusted authority (often referred to as the *certification authority*). The result of this endorsement is called a *certificate.* A certificate is created when an entity's credentials are signed by the certification authority and is used for proof of an entity's identity to other parties.

combination card A smart card with both contact and contactless features.

contactless card An integrated circuit card that enables energy to flow between the card and the interfacing device without the use of contact. Instead, induction or high-frequency transmission techniques are used through a radio frequency (RF) interface.

cryptographic algorithm The method for providing information security services. DES (data encryption standard) is the method most commonly used to encrypt and decrypt information.

cryptography The study of providing information security services.

data integrity   The accurate and timely delivery of information.

Decryption   The process of transforming encrypted information (ciphertext) into plain text.

DES (data encryption standard)   The most popular commercial symmetric key encryption system.

digital signature   A means to bind information to the identity of the originator or owner of that information. Digital signatures provide (1) origin (entity) authentication, (2) data integrity, and (3) signer nonrepudiation.

ECC (elliptic curve cryptosystem)   A public key cryptosystem that provides higher cryptographic strength using smaller key sizes than other public key cryptosystems. For example, an ECC 160-bit key is stronger than a 1024-bit RSA key.

EEPROM   An acronym for electronically erasable programmable read-only memory. A nonvolatile memory technology that allows data to be electrically erased and rewritten.

encryption   The process of transforming plain text into ciphertext for confidentiality or privacy.

entity authentication   A means to identify an entity (an entity might be a person or persons, a computer terminal, a credit card, a fax, etc.).

key   A value used in encryption and digital signatures to provide the information security services.

memory card   An integrated circuit card capable of storing information but not having calculating capability, i.e., no microprocessor.

microprocessor card   A microcomputer with all of its processing facilities on a single chip. Also called *microprocessor-on-a-chip.* A microprocessor is a computer processor on a chip, including registers and a possible cache memory. A microcomputer or microcontroller also has data and program memory on the same chip.

nonrepudiation   An information security service that provides irrefutable proof of the identity of the signer and the integrity of the data. This can be verified by a third party.

private key   In a public key system, it is that key in a key pair that is held by the individual entity and never revealed. The private key is embedded in the product, and a hardware platform is preferable for containing this critical piece of data.

public key   In a public key system, it is that key in a key pair that is publicized.

public key encryption   Also known as *asymmetric key encryption,* this is a cryptographic algorithm that uses different keys for encryption (e) and decryption (d), where (e) and (d) are mathematically linked. It is computationally infeasible to determine (d) from (e). Therefore, this system allows the distribution of (e) (public key) without disclosing (d) (private key). Public key cryptography is the most important advancement in the field of cryptography in the last 2000 years.

random access memory (RAM)   A volatile memory used in integrated circuit cards that requires power to maintain data.

read-only memory (ROM)   Nonvolatile memory that is written once, usually during card production. It is used to store operating systems and algorithms employed by the microprocessor in an integrated circuit card during transactions.

RSA   A commonly used public key encryption system that must be licensed for use in the United States. Since current implementations of RSA are relatively slow, it is often used for challenge and authentication, and a symmetric key encryption system such as DES is used to encrypt information.

symmetric key encryption   In this system, the same key is used to encrypt and decrypt text. Key management is less efficient in this encryption system than in public key systems.

The Smart Manager's Decision Checklists

Application and System Decisions

If the Application Requires	Then Use
a. Low cost	Disposable memory cards
b. Memory	
A great deal of memory	EEPROM, RAM, Flash
A rewritable memory	EEPROM, RAM, Flash
A small amount of memory	ROM
b. Communications protocols	
Synchronous	Memory cards
Asynchronous	Microcontroller cards
c. Processing capability	
Fast authentication of security keys	Coprocessor
Complicated algorithm(s)	Coprocessor
Speed	Single processor
PIN checking	Single processor/memory
None	Memory
c. Silicon size	
Cost	Vendor dependent
d. Operating system	
7816 compliant	Buy
Closed or proprietary	Build and support

If the Application Requires	Then Use
e. Mask	
7816 compliant	Buy
Closed or proprietary	Build and support
Multiple versions	Build and support
f. File structure	
Quick access	Hierarchical
Common shared data	Relational
Unstructured	Object-oriented
g. Security	Always
h. Processing speed	
ISO compliant	3 to 5 MHz
All others	Terminal dependent
Power (1, 3, 5 volts)	Reader/card interdependent
Clock speed	Power dependent
i. Testing	Mandatory
j. Certification	As needed
k. Contact/contactless cards	Application dependent
l. Terminals	
Accept memory cards	Always plan for
Accept microcontroller cards	Always plan for

Manufacturing Decisions

If the Application Requires	Then Use
a. Long life/lamination	Sheet offset
b. High-volume print runs	Sheet offset
c. Low-volume/low-quality print runs	Injection molding
d. High-quality card	Sheet offset
e. Consistent cavity	Injection molding
f. Disposable cards	Injection molding

System Implementation Decisions

If the Application Requires	Then Use
a. Communications	
High speed	Contact cards
Low speed	Contactless cards
Remote	Contactless cards
b. Single or multiple applications	Memory, ROM, OS dependent
c. Fraud protection	Crypto coprocessor
d. High level of security	Crypto coprocessor
e. Amount of memory	Application(s) dependent
f. Single or multiple cards used	Application(s) dependent
g. Ability to detect failures	Semiconductor, OS, application(s) dependent
h. Terminal types	Application(s) dependent
i. Compliant with standards	Always

Functional Design Decisions

If the Application Requires	Then Use
a. Long registers	Semiconductor size dependent
b. High security	Coprocessor
c. High-speed communications	Non-ISO clock speeds
d. Storage registers	Memory dependent
e. Anticounterfeit	Tamper evident/resistent
f. Volatile versus nonvolatile memory	Application(s) dependent

Importance of Standards and Specifications—Smart Manager

There are a number of standards and specifications that the smart manager should understand in detail before undertaking a smart card–based project. Some specifications and standards may be focused on a single industry. The EMV specification introduced in Chapter 3 is focused on the financial card sector. Other standards, such as ISO and CEN standards, cover the entire spectrum of card applications.

ISO has evolved standards (note the word *evolved*) over the past 10 years to describe an open and international common format for smart cards (contactless cards are evolving as this book is being published).

Most standards are modular—broken into specific pieces that describe the physical, logical, and interactive characteristics. In being standards-compliant, a useful question to ask is to what level (e.g., ISO 7816, Parts 3, 4, 5, 6).

The smart manager must appreciate that these standards are evolving. As with most high technologies, there are new concepts and ideas incorporated into solutions each and every day. As ISO standards evolve, there are a few areas that the smart manager should give extra attention to.

Evolution of smart cards in the future will most likely include the following:

◆ Lower-voltage cards

◆ Higher-frequency (faster) cards

◆ Full-duplex transmission (today's cards are half duplex)

◆ Higher-speed encryption/decryption

As new technologies are implemented, these and other concepts may appear in "reissued" standards and specifications. Both standards and specifications have "sunset" provisions that require periodic reviews to renew, modify, retire—or add as needed—additional modules (levels).

The EMV specification is in four parts. We have included excerpts of the EMV specification in this appendix to provide an introductory understanding of standards and specifications. Note that these are only excerpts, not complete copies, and the information should not be used out of their full context.

EMV is divided into four parts:

◆ Physical, logical, and transmission (parallel to ISO 7816, parts 1 to 3)

◆ Data Elements and Commands

◆ Application Selection

◆ Security Aspects

Part I, for example, shows the specific contacts and pin assignments that are to be used by EMV-compliant smart cards. These specifications are more specific than ISO 7816 and describe the exact manner in which a card must be used to support financial transaction applications.

EMV introduces the concept of data elements and files as a specific requirement to be used by the smart card. Examples in Part II are reproduced to demonstrate this point. This includes classification of objects and "tagging" of objects that would be used in EMV (and nonfinancial) uses.

Part III describes the way in which applications (as it is envisioned that more than one application would be supported in a smart card) are selected by both the smart card and terminal. This follows the ISO 7816 part 5 concepts, but provides for specific assignments of applications and specific identifiers to be used in financial transactions.

Part IV describes the security aspects that shall be used in financial transaction applications. There are two types of authentication—static and dynamic. Both are specified by EMV.

Representative annexes to the specification are also included to demonstrate the types of data elements that may be found in the specification and examples of "EMV Tags" that would be applied in financial transaction applications.

The full EMV specification (and ongoing enhancements) can be found on the Internet at a number of sites, including www.visa.com and www.mastercard.com

Understanding and keeping track of current standards and specifications and monitoring for changes and evolution is an important part of the smart manager's day.

EMV '96 Integrated Circuit Card Specification for Payment Systems

Version 3.0
June 30, 1996

Table of Contents

1. Scope

The *Integrated Circuit Card (ICC) Specification for Payment Systems* describes the minimum functionality required of integrated circuit cards (ICCs) and terminals to ensure correct operation and interoperability. Additional proprietary functionality and features may be provided, but these are beyond the scope of this specification and interoperability cannot be guaranteed.

This specification consists of four parts:

> *Part I—Electromechanical Characteristics, Logical Interface, and Transmission Protocols*
>
> *Part II—Data Elements and Commands*
>
> *Part III—Application Selection*
>
> *Part IV—Security Aspects*

Part I defines electromechanical characteristics, logical interface, and transmission protocols as they apply to the exchange of information between an ICC and a terminal. In particular it covers:

- Mechanical characteristics, voltage levels, and signal parameters as they apply to both ICCs and terminals.
- An overview of the card session.
- Establishment of communication between the ICC and the terminal by means of the answer to reset.
- Character- and block-oriented asynchronous transmission protocols.

Part II defines data elements and commands as they apply to the exchange of information between an ICC and a terminal. In particular it covers:

- Data elements for financial interchange and their mapping onto data objects.
- Structure and referencing of files.

◆ Structure and coding of messages between the ICC and the terminal to achieve application level functions.

Part III defines the application selection process from the standpoint of both the card and the terminal. The logical structure of data and files within the card that is required for the process is specified, as is the terminal logic using the card structure.

Part IV defines the security aspects of the processes specified in this specification. In particular it covers:

◆ Static data authentication.

◆ Dynamic data authentication.

◆ Secure messaging.

This specification does not cover the details of Transaction Certificate generation by the ICC, the internal implementation in the ICC, its security architecture, and its personalisation.

This specification is based on the ISO/IEC 7816 series of standards and should be read in conjunction with those standards. However, if any of the provisions or definitions in this specification differ from those standards, the provisions herein shall take precedence.

This specification is intended for a target audience that includes manufacturers of ICCs and terminals, system designers in payment systems, and financial institution staff responsible for implementing financial applications in ICCs.

2. Normative References

The following standards contain provisions that are referenced in this specification.

Europay, MasterCard, and Visa (EMV): June 30, 1996	Integrated Circuit Card Application Specification for Payment Systems
Europay, MasterCard, and Visa (EMV): June 30, 1996	Integrated Circuit Card Terminal Specification for Payment Systems
FIPS Pub 180-1:1995	Secure Hash Standard
IEC 512-2:1979	Specifications for electromechanical components for electromechanical equipment—Part 2: Contact resistance tests, insulation tests, and voltage stress tests.
ISO 639:1988	Codes for the representation of names and languages
ISO 3166:1993	Codes for the representation of names and countries
ISO 4217:1990	Codes for the representation of currencies and funds
ISO/IEC 7811-1:1992	Identification cards—Recording technique—Part 1: Embossing
ISO/IEC 7811-3:1992	Identification cards—Recording technique—Part 3: Location of embossed characters on ID-1 cards
ISO/IEC 7813:1990	Identification cards—Financial transaction cards
ISO 7816-1:1987	Identification cards—Integrated circuit(s) cards with contacts—Part 1: Physical characteristics
ISO 7816-2:1988	Identification cards—Integrated circuit(s) cards with contacts—Part 2: Dimensions and location of contacts
ISO/IEC 7816-3:1989	Identification cards—Integrated circuit(s) cards with contacts—Part 3: Electronic signals and transmission protocols
ISO/IEC 7816-3:1992	Identification cards—Integrated circuit(s) cards with contacts—Part 3, Amendment 1: Protocol type T=1, asynchronous half duplex block transmission protocol

ISO/IEC 7816-3:1994	Identification cards—Integrated circuit(s) cards with contacts—Part 3, Amendment 2: Protocol type selection (Draft International Standard)
ISO/IEC 7816-4:1995	Identification cards—Integrated circuit(s) cards with contacts—Part 4, Inter-industry commands for interchange
ISO/IEC 7816-5:1994	Identification cards—Integrated circuit(s) cards with contacts—Part 5: Numbering system and registration procedure for application identifiers
ISO/IEC 7816-6:1995	Identification cards—Integrated circuit(s) cards with contacts—Part 6: Inter-industry data elements (Draft International Standard)
ISO 8731-1:1987	Banking—Approved algorithms for message authentication—Part 1: DEA
ISO 8372:1987	Information processing—Modes of operation for a 64-bit block cipher algorithm
ISO/IEC 8825:1990	Information technology—Open systems inter connection—Specification of basic encoding rules for abstract syntax notation one (ASN.1)
ISO 8583:1987	Bank card originated messages—Interchange message specifications—Content for financial transactions
ISO 8583:1993	Financial transaction card originated messages—Interchange message specifications
ISO 8859:1987	Information processing—8-bit single-byte coded graphic character sets
ISO/IEC CD 9796-2: 1996	Information technology—Security techniques—Digital signature scheme giving message recovery—Part 2: Mechanism using a hash function
ISO/IEC 9797:1993	Information technology—Security techniques—Data integrity mechanism using a cryptographic check function employing a block cipher algorithm
ISO/IEC 10116:1993	Information technology—Modes of operation of an n-bit block cipher algorithm
ISO/IEC CD 10118-3: 1996	Information technology—Security techniques—Hash functions—Part 3: Dedicated hash functions
ISO/IEC 10373:1993	Identification cards—Test methods

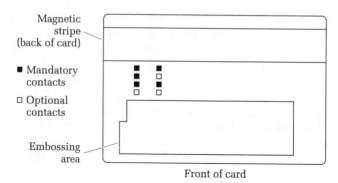

Figure I-1   Location of Contacts

1.1.3 Contact Assignment

The assignment of the ICC contacts shall be as defined in ISO 7816-2 and is shown in Table I-1:

TABLE I-1   ICC Contact Assignment

C1	Supply voltage (VCC)	C5	Ground (GND)
C2	Reset (RST)	C6	Not used[1]
C3	Clock (CLK)	C7	Input/output (I/O)

C4 and C8 are not used and need not be physically present. C6 is not used and need not be physically present; if present, it shall be electrically isolated[2] from the integrated circuit (IC) itself and other contacts on the ICC.

1.2 Electrical Characteristics of the ICC

This section describes the electrical characteristics of the signals as measured at the ICC contacts.

---

[1] Defined in ISO/IEC 7816 as programming voltage (VPP).
[2] Electrically isolated means that the resistance measured between C6 and any other contact shall be $\geq 10 M\Omega$ with an applied voltage of 5V DC.

1.2.1 Measurement Conventions

All measurements are made at the point of contact between the ICC and the interface device (IFD) contacts and are defined with respect to the GND contact over an ambient temperature range 0° C to 50° C.

All currents flowing into the ICC are considered positive.

2. Card Session

This section describes all stages involved in a card session from insertion of the ICC into the IFD through the execution of the transaction to the removal of the ICC from the IFD.

2.1 Normal Card Session

This section describes the processes involved in the execution of a normal transaction.

2.1.1 Stages of a Card Session

A card session is comprised of the following stages:

1. Insertion of the ICC into the IFD and connection and activation of the contacts.
2. Reset of the ICC and establishment of communication between the terminal and the ICC.
3. Execution of the transaction(s).
4. Deactivation of the contacts and removal of the ICC.

2.1.2 ICC Insertion and Contact Activation Sequence

On insertion of the ICC into the IFD, the terminal shall ensure that all signal contacts are in state L with values of V_{OL} as defined in section I-1.4 and that V_{CC} is 0.4 V or less before any contacts are physically made. The IFD shall be able to detect when the ICC is seated to within ±0.5 mm of the nominally correct position[5] in the direction of insertion/withdrawal. When the IFD detects that the ICC is seated within this tolerance, and when all contacts have been physically made, the contacts shall be activated as follows (see Figure I-2):

---

[5]The 'nominally correct position' is when the centres of the IFD contacts are exactly over the centres of the ICC contacts located as specified in ISO 7816-2.

- ◆ RST shall be maintained by the terminal in state L throughout the activation sequence.
- ◆ Following establishment of the physical contacts but prior to activation of I/O or CLK, VCC shall be powered.

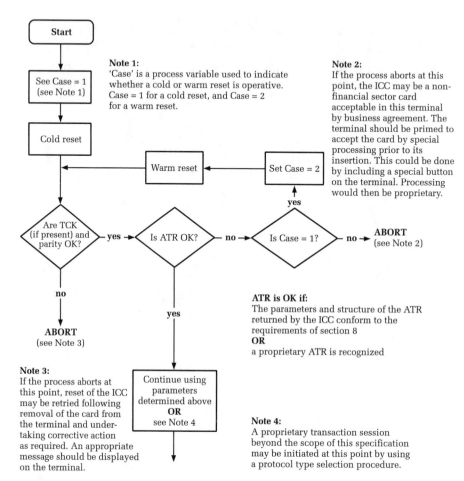

Note 1:
'Case' is a process variable used to indicate whether a cold or warm reset is operative. Case = 1 for a cold reset, and Case = 2 for a warm reset.

Note 2:
If the process aborts at this point, the ICC may be a non-financial sector card acceptable in this terminal by business agreement. The terminal should be primed to accept the card by special processing prior to its insertion. This could be done by including a special button on the terminal. Processing would then be proprietary.

ATR is OK if:
The parameters and structure of the ATR returned by the ICC conform to the requirements of section 8
OR
a proprietary ATR is recognized

Note 3:
If the process aborts at this point, reset of the ICC may be retried following removal of the card from the terminal and undertaking corrective action as required. An appropriate message should be displayed on the terminal.

Note 4:
A proprietary transaction session beyond the scope of this specification may be initiated at this point by using a protocol type selection procedure.

Figure I-7   ATR—Example Flow at the Terminal

1. Data Elements and Files

An application in the ICC includes a set of items of information. These items of information may be accessible to the terminal after a successful application selection (see Part III of this specification).

An item of information is called a data element. A data element is the smallest piece of information that may be identified by a name, a description of logical content, a format, and a coding.

1.1 Data Elements Associated with Financial Transaction Interchange

The data element directory defined in Annex B, Table B-1 includes those data elements that may be used for financial transaction interchange. Data elements not specified in Annex B, Table B-1, are outside the scope of these specifications.

1.2 Data Objects

A data object consists of a tag, a length, and a value. A tag uniquely identifies a data object within the environment of an application. The length is the length of the value field of the data object. The value of a data object may consist either of a data element or of one or more data objects. When a data object encapsulates a data element, it is called a primitive data object. When a data object encapsulates one or more data objects, it is called a constructed data object. Specific tags are assigned to the constructed data objects with a specific meaning in the environment of an application according to this specification. The value field of such constructed data objects is a context-specific template. Rules for the coding of context-specific data objects and templates are given in Annex C.

Table B-1 in Annex B describes the mapping of data elements onto data objects and the mapping of data objects into templates when applicable.

Records are templates containing one or more primitive and/or constructed data objects.

The mapping of data objects into records is left to the discretion of the issuer and the manner in which data elements are to be used is described in the *ICC Application Specification for Payment Systems.*

1.2.1 Classes of Data Objects

Identification and coding of different classes of data objects are defined in Annex C. The tag definitions specified in that annex are according to ISO/IEC 8825 and ISO/IEC 7816 series and apply to applications conforming to this specification.

1. Application Selection

This section describes the application selection process from the standpoint of both the card and the terminal. The logical structure of data and files within the card that are required for the process is specified, after which the terminal logic using the card structure is described.

The application selection process described in this section is the process by which the terminal uses data in the ICC according to protocols defined herein to determine which payment system application is to be run for a transaction. The process is described in two steps:

1. Create a list of applications that are mutually supported by the card and the terminal.
2. Select the application to be run from the list generated by step 1.

It is the intent of this section to describe the necessary information in the card and two terminal selection algorithms that yields the correct results. Other terminal selection algorithms that yield the same results are permitted in place of the selection algorithms described here.

Application selection is always the first application function performed.

A payment system application is comprised of the following:

- ◆ A set of files in the ICC providing data customised by the issuer.
- ◆ Data in the terminal provided by the acquirer or the merchant.
- ◆ An application protocol agreed upon by both the ICC and the terminal.

Applications are uniquely identified by AIDs conforming to ISO/IEC 7816-5 (see section III-1.1).

The techniques chosen by the payment systems and described herein are designed to meet the following key objectives:

◆ Ability to work with ICCs with a wide range of capabilities.

◆ Ability for terminals with a wide range of capabilities to work with all ICCs supporting payment system applications according to this specification.

◆ Conformance with ISO standards.

◆ Ability of ICCs to support multiple applications, not all of which need to be payment system applications.

◆ As far as possible, provide the capability for applications conforming with this specification to co-reside on cards with presently existing applications.

◆ Minimum overhead in storage and processing.

◆ Ability for the issuer to optimise the selection process.

The set of data that the ICC contains in support of a given application is defined by an ADF selected by the terminal using a SELECT command and an AFL returned by the ICC in response to a GET PROCESSING OPTIONS command.

1.1 Coding of Payment System Application Identifier

The structure of the AID is according to ISO/IEC 7816-5 and consists of two parts:

1. A Registered Application Provider Identifier (RID) of 5 bytes, unique to an application provider and assigned according to ISO/IEC 7816-5.

2. An optional field assigned by the application provider of up to 11 bytes. This field is known as a Proprietary Application Identifier Extension (PIX) and may contain any 0–11 byte value specified by the provider. The meaning of

this field is defined only for the specific RID and need not be unique across different RIDs.

Additional ADFs defined under the control of other application providers may be present in the ICC but shall avoid duplicating the range of RIDs assigned to payment systems. Compliance with ISO/IEC 7816-5 shall assure this avoidance.

1.2 Structure of the Payment Systems Environment

The Payment Systems Environment begins with a Directory Definition File (DDF) given the name '1PAY.SYS.DDF01'. The presence of this DDF is mandatory. This DDF is mapped onto a DF within the card, which may or may not be the MF. As with all DDFs, this DDF shall contain a Payment Systems Directory. The FCI of this DDF shall contain at least the information defined for all DDFs in Part II, and, optionally, the Language Preference (tag '5F2D') and the Issuer Code Table Index (tag '9F11').

The directory attached to this initial DDF contains entries for ADFs that are formatted according to this specification, although the applications defined by those ADFs may or may not conform to this specification. The directory may also contain entries for other Payment System's DDFs, which shall conform to this specification.

The directory is not required to have entries for all DDFs and ADFs in the card, and following the chain of DDFs may not reveal all applications supported by the card. However, only applications that are revealed by following the chain of DDFs beginning with the initial directory can be assured of international interoperability.

1. Static Data Authentication

Static data authentication is performed by the terminal using a digital signature based on public key techniques to confirm the legitimacy of critical ICC-resident static data identified by the AFL. This detects unauthorised alteration of data after personalisation.

Static data authentication requires the existence of a certification authority, which is a highly secure cryptographic facility that 'signs' the issuer's public keys. Every terminal conforming to this specification shall contain the appropriate certification authority's public key(s) for every application recognised by the terminal. This specification permits multiple AIDs to share the same 'set' of certification authority public keys. The relationship between the data and the cryptographic keys is shown in Figure IV-1.

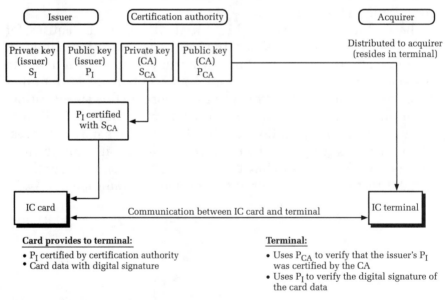

Figure IV-1   Diagram of Static Data Authentication

ICCs that support static data authentication shall contain the following data elements:

◆ Certification Authority Public Key Index: This one-byte data element contains a binary number that indicates which of the application's certification authority public keys and its associated algorithm that reside in the terminal is to be used with this ICC.

◆ Issuer Public Key Certificate: This variable-length data element is provided by the appropriate certification authority to the card issuer. When the terminal verifies this data element, it authenticates the Issuer Public Key plus additional data as described in section IV-1.3.

◆ Signed Static Application Data: A variable-length data element generated by the issuer using the private key that corresponds to the public key authenticated in the Issuer Public Key Certificate. It is a digital signature covering critical ICC-resident static data elements, as described in section IV-1.4.

◆ Issuer Public Key Remainder: A variable length data element. Its presence in the ICC is optional. See section IV-1.1 for further explanation.

◆ Issuer Public Key Exponent: A variable length data element provided by the issuer. See section IV-1.1 for further explanation.

To support static data authentication, each terminal shall be able to store multiple certification authority public keys and shall associate with each such key the key-related information to be used with the key (so that terminals can in the future support multiple algorithms and allow an evolutionary transition from one to another). The terminal shall be able to locate any such key (and the key-related information) given the RID and Certification Authority Public Key Index as provided by the ICC.

Static data authentication shall use a reversible algorithm as specified in Annex E2.1 and Annex F2. Section IV-1.1 contains an overview of the keys and certificates involved in the static data authentication process, and sections IV-1.2 to IV-1.4 specify the three main steps in the process, namely

- ◆ Retrieval of the Certification Authority Public Key by the terminal.
- ◆ Retrieval of the Issuer Public Key by the terminal.
- ◆ Verification of the Signed Static Application Data by the terminal.

1.1 Keys and Certificates

To support static data authentication, an ICC shall contain the Signed Static Application Data, which is signed with the Issuer Private Key. The Issuer Public Key shall be stored on the ICC with a public key certificate.

The bit length of all moduli shall be a multiple of 8, the leftmost bit of its leftmost byte being 1. All lengths are given in bytes.

The signature scheme specified in Annex E2.1 is applied to the data specified in Table IV-1 using the Certification Authority Private Key S_{CA} in order to obtain the Issuer Public Key Certificate.

2. Dynamic Data Authentication

Dynamic data authentication is performed by the terminal using a digital signature based on public key techniques to authenticate the ICC, and confirm the legitimacy of critical ICC-resident data identified by the ICC dynamic data and data received from the terminal identified by the Dynamic Data Authentication Data Object List (DDOL). This precludes the counterfeiting of any such card.

Dynamic data authentication requires the existence of a certification authority, a highly secure cryptographic facility that 'signs' the Issuer's Public Keys. Every terminal conforming to this specification shall contain the appropriate certification authority's public key(s) for every application recognised by the terminal. This specification permits multiple AIDs to share the same 'set' of certification authority public keys. The relationship between the data and the cryptographic keys is shown in Figure IV-2.

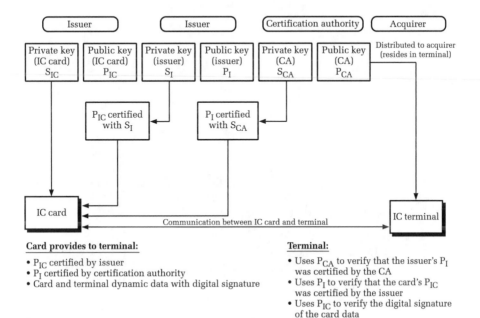

Figure IV-2   Diagram of Dynamic Data Authentication

ICCs that support dynamic data authentication shall contain the following data elements:

♦ Certification Authority Public Key Index: This one-byte data element contains a binary number that indicates which of the application's certification authority

Annex B—Data Elements Table

Table B-1 defines those data elements that may be used for financial transaction interchange and their mapping onto data objects and files.

Name	Description	Source	Format	Template	Tag	Length
Acquirer Identifier	Uniquely identifies the acquirer within each payment system	Terminal	n 6–11	—	'9F01'	6
Additional Terminal Capabilities	Indicates the data input and output capabilities of the terminal	Terminal	b	—	'9F40'	5
Amount, Authorised (Binary)	Authorised amount of the transaction (excluding adjustments)	Terminal	b	—	'81'	4
Amount, Authorised (Numeric)	Authorised amount of the transaction (excluding adjustments)	Terminal	n 12	—	'9F02'	6
Amount, Other (Binary)	Secondary amount associated with the transaction representing a cashback amount	Terminal	b	—	'9F04'	4
Amount, Other (Numeric)	Secondary amount associated with the transaction representing a cashback amount	Terminal	n 12	—	'9F03'	6
Amount, Reference Currency	Authorised amount expressed in the reference currency	Terminal	b	—	'9F3A'	4
Application Cryptogram	Cryptogram returned by the ICC in response of the GENERATE AC command	ICC	b	'79' or '80'	'9F26'	8
Application Currency Code	Indicates the currency in which the account is managed according to ISO 4217	ICC	n 3	'70' or '79'	'9F42'	2
Application Currency Exponent	Indicates the implied position of the decimal point from the right of the account represented according to ISO 4217	ICC	n 1	'70' or '79'	'9F44'	1
Application Discretionary Data	Issuer or payment system specified data relating to the application	ICC	b	'70' or '79'	'9F05'	1–32

The tags allocated to the data elements are according to Table B-2:

Name	Template	Tag
Application Identifier (AID)	'61'	'4F'
Application Label	'61'	'50'
Command to Perform	'61'	'52'
Track 2 Equivalent Data	'70' or '79'	'57'
Application Primary Account Number (PAN)	'70' or '79'	'5A'
Cardholder Name	'70' or '79'	'5F20'
Application Expiration Date	'70' or '79'	'5F24'
Application Effective Date	'70' or '79'	'5F25'
Issuer Country Code	'70' or '79'	'5F28'
Transaction Currency Code	—	'5F2A'
Language Preference	'A5'	'5F2D'
Service Code	'70' or '79'	'5F30'
Application Primary Account Number (PAN) Sequence Number	'70' or '79'	'5F34'
Transaction Currency Exponent	—	'5F36'
Application Template	'70' or '79'	'61'
File Control Information (FCI) Template	—	'6F'
Application Elementary File (AEF) Data Template	—	'70'
Issuer Script Template 1	—	'71'
Issuer Script Template 2	—	'72'
Directory Discretionary Template	'61'	'73'
Response Message Template Format 2	—	'79'
Response Message Template Format 1	—	'80'
Amount, Authorised (Binary)	—	'81'
Application Interchange Profile	'79' or '80'	'82'
Command Template	—	'83'
Dedicated File (DF) Name	'6F'	'84'
Issuer Script Command	'71' or '72'	'86'
Application Priority Indicator	'61' or 'A5'	'87'
Short File Identifier (SFI)	'A5'	'88'
Authorisation Code	—	'89'
Authorisation Response Code	—	'8A'
Card Risk Management Data Object List 1 (CDOL1)	'70' or '79'	'8C'
Card Risk Management Data Object List 2 (CDOL2)	'70' or '79'	'8D'
Cardholder Verification Method (CVM) List	'70' or '79'	'8E'

Name	Template	Tag
Certification Authority Public Key Index	'70' or '79'	'8F'
Issuer Public Key Certificate	'70' or '79'	'90'
Issuer Authentication Data	—	'91'
Issuer Public Key Remainder	'70' or '79'	'92'
Signed Static Application Data	'70' or '79'	'93'
Application File Locator (AFL)	'79' or '80'	'94'
Terminal Verification Results	—	'95'
Transaction Certificate Data Object List (TDOL)	'70' or '79'	'97'
Transaction Certificate (TC) Hash Value	—	'98'
Transaction Personal Identification Number (PIN) Data	—	'99'
Transaction Date	—	'9A'

ENCRYPTION
TECHNOLOGIES

Cryptography, the writing and deciphering of messages in secret code, has evolved from a rich past. Cryptography has been used for centuries to protect sensitive information as it is transmitted from one location to another. Its purpose has expanded from delivering messages across enemy lines (decoder rings) to include providing a safe environment for electronic transmissions.

Encryption is the process of writing the message and using a predetermined key to create a secret message. Decryption is the reverse of this process. In general terms, all cryptographic systems work in this manner. The only difference may be in the details of the process or the algorithm.

This appendix describes specific encryption techniques and their mathematical algorithms. There are two primary encryption methods in use today:

1. Secret key cryptography
2. Public key cryptography

Secret Key Cryptosystems

Secret key cryptography, also known as *symmetric cryptography,* uses the same key to encrypt and decrypt the message. The sender

Before we begin our discussion of cryptography, though, here are a few terms that you should be familiar with.

application   The application protocol between the card and the terminal and its related set of data.

asymmetric cryptographic algorithm   A cryptographic technique that uses two related transformations, a public key and a private transformation that is defined by the private key. The two transformations have the property that, given the public transformation, it is computationally not feasible to derive the private transformation.

ciphertext   Encrypted information.

cryptogram   Result of a cryptographic operation.

cryptographic algorithm   An algorithm that transforms data in order to hide or reveal its information content.

data integrity   The property that data has not been altered or destroyed in an unauthorized manner.

digital signature   The process of calculating, and consequently verifying, a digital signature. This process requires a message to be passed through a hash function, thus producing a short block, and then verified against forgery and data integrity.

function   A process accomplished by one or more commands and resultant actions that are used to perform all or part of a transaction.

hash function   A function that maps strings of bits to fixed-length strings of bits. A hash is always computationally not feasible to allow for a given output an input that maps to this point.

hash result   The string of bits that is the output of a hash function.

Continued

key   A sequence of symbols that controls the operation of a cryptographic transformation.

message authentication code   A symmetric cryptographic transformation of data that protects the sender and the recipient of the data against forgery by third parties.

padding   Appending extra bits to either side of a data string.

private key   Part of the asymmetric key pair that should be used only by that entity. In the case of a digital signature, the private key defines the signature function.

public key   Part of the asymmetric key pair that can be made public. In the case of a digital signature, the public key defines the verification function.

symmetric cryptographic algorithm   A cryptographic technique that uses the same secret key for both the originator's and recipient's transformation without knowledge of the secret key, it is computationally not feasible to compute either the originator's or the recipient's transformation.

and recipient of a message must share a secret, namely the key. A well-known secret key cryptography algorithm is the Data Encryption Standard (DES), which is used by financial institutions to encrypt PINs.

Public Key Cryptosystems

Public key cryptography, also known as *asymmetric cryptography,* uses two keys: one key to encrypt the message and the other key to decrypt the message. The two keys are mathematically related such that data encrypted with either key can be decrypted only by using the other. A user has two keys—a public key and a private key—and is responsible for distributing the public key.

Because of the relationship between the two keys, the user and anyone receiving the public key can be assured that data encrypted with the public key and sent back to the user can be decrypted by the receiver using the private key. This assurance of secrecy, though, is maintained only if the user ensures that the private key is not disclosed to another person. The best-known public key cryptography algorithm is RSA (named after its inventors Rivest, Shamir, and Adleman).

General Description of a Public and Private Key Algorithm

The following is for illustrative purposes only.

Stage 1: Select key length.

Equate two large prime numbers.

Defined as A and B.

1,024 bits in length.

Stage 2: Computation.

Select C such that C is less than AB, and such that C and $(A - 1)(B - 1)$ maintain no prime factors in common.

$C < AB$.

Stage 3: Computation.

Compute D such that $(DC - 1)$ is divisible by $(A - 1)(B - 1)$ with no remainder. (This factoring relies on the Chinese remainder theorem [CRT], where

$$M_1 = x^s \pmod{y} \quad \text{and} \quad M_2 = x^s \pmod{z}$$
$$M = x^s \pmod{N}$$

This is the heart of the security, where factoring the modulus Y is difficult and a unique solution modulus is created. If the factors of an integer are known or easily computed, then someone could easily break the system.)

$DC = 1 \mod (A - 1)(B - 1)$.

D is the multiplicative inverse of C.

Stage 4: Encryption.

E equals E raised to the power of F, where E is a positive integer.

Encrypt $(E) = (E \wedge F) \mod AB$.

Stage 5: Decryption.

G is a positive integer, also known as the ciphertext.

Decrypt $(G) = (G \wedge D) \mod AB$.

Stage 6: Identify keys.

The product AB is the modulus.

The public key is the pair (AB, C).

The private key is the number D.

The public key is open and may be published on the Internet or given to friends and acquaintances without compromising your security. The reason public keys can be given freely is because it is nearly impossible (or at least resource unfeasible) to calculate D, A, or B given only the public key (AB, C).

Secret key–only cryptography is impractical for exchanging messages with a large group of previously unknown correspondents over a public network. In order for a merchant to conduct transactions securely with millions of Internet subscribers, each consumer would need a distinct key assigned by the merchant and transmitted over a separate secure channel. On the other hand, by using public key cryptography, that same merchant could create a public/private key pair and publish the public key, allowing any consumer to send a secure message to the merchant.

Remember, A and B can be of any length, but 1,024 bits provides the most security. When this is the case, the most powerful supercomputers of today cannot complete the computationally heavy task of factoring, thus providing the security required from the key.

(The reason key lengths are not longer is because of U.S. government regulations. The government maintains export restrictions and monitors keys and key lengths.)

The algorithm is such that changing a single bit in the message will on average change half of the bits in the encrypted message. The odds of two messages having the same encryption are about one in 1,000,000,000,000,000,000,000,000,000,000,000,000,000,000,000! As you can see, it is computationally unfeasible to generate two different messages that have the same encrypted form.

Digital Signature

Integrity and authentication are ensured by the use of digital signatures, which provide the same function as handwritten signatures. A digital signature is a piece of electronic data asserting that a named individual created or agreed to authorize the accompanying information. With digital signatures, the recipient or third party can also verify that both the document and the signature have not been altered since they were sent. A secure digital signature system thus consists of two parts: (1) a method of signing a document such that forgery is not feasible and (2) a method of verifying that a signature was actually generated by whomever it represents. Furthermore, secure digital signatures cannot be repudiated; the signer cannot refuse ownership at a later period.

There is no required relationship between public and private keys and digital signatures. Each is a separate security method enhanced by its coupling with the other. When combined with message digests, encryption using the private key allows users to digitally sign messages. A message digest is a value generated for a message or document that is unique to that message. A message digest is generated by passing the message through a one-way cryptographic function (i.e., one that cannot be reversed). When the digest of a message is encrypted using the sender's private key and is appended to the original message, the result is known as the *digital signature* of the message.

Cryptographic Attacks

The smart manager will recognize that cryptographic attacks may come from compromising people, not just the hardware. Proper restrictions and background checks, along with accounting systems, are required for key management. There are five main types of cryptographic attacks:

- ◆ Exhaustive key search
- ◆ Intentional message corruption
- ◆ Corruption of internal data
- ◆ Direct manipulation
- ◆ External attacks

Exhaustive Key Search

Cryptosystems require an encryption key or keys. For each given algorithm there are only a finite number of possible keys that can be selected. This subset may be large and comprehensive; however, it is still limited in scope. With any limited set, it is possible to complete an exhaustive key search.

Exhaustive key search, also known as a *brute-force technique*, is analogous to finding a needle in a haystack. With a supercomputer and no concept of time, it is possible to test every possible key in turn until the actual one being used is found.

The smart manager will protect the system by selecting the longest possible key length, thus making it more difficult to conduct a search. In essence, given today's resources, it is unfeasible to complete an exhaustive search. Technology does not stand still, and the smart manager will update the system as computing power increases.

Intentional Message Corruption

In this process an attacker will attempt to alter the message and affect the output of the data. The system must detect the outside interference and act upon the attack. The smart manager will design a system that documents when an intentional message corruption is committed.

The smart manager can protect against this attack by the use of digital signatures. Digital signatures offer sufficient protection to guard against any interference that is carried out externally.

Corruption of Internal Data

This attack is an intricate process and allows for retrograde information only. The corruption of internal data is completed by deduction methods. In other words, the attacker introduces single-bit errors in the processor performing the encryption. By monitoring the results of the error, a person could deduce information concerning the encryption algorithm and the keys. This method is based on theory. The information that an attacker could obtain is open to question.

Direct Manipulation

If an attacker is able to compromise the processor itself, then it is possible to directly manipulate the internal operations of the processor and alter the data. This attack is highly unlikely since the operating system, the cryptosystem, and the silicon all incorporate various preventive measures.

External Attacks

Although external attacks are guarded by certain barriers, it is important to know the points of convergence. The power supply, clock, reset function, and I/O-bidirectional data line are all contact connections and potential areas of concern.

Most Commonly Used Commercial Cryptosystems

DES

Data Encryption Standard (DES) was developed by IBM in the late 1970s and evolved into the official encryption standard of the U.S.

government. The export of DES is regulated by the U.S. government and even defined in an ANSI standard. Generally, DES is considered unbreakable with the system of secret keys designed to be implemented in the hardware.

The only known method of breaking DES is the manipulation of all possible key lengths and combinations. This is called an *exhaustive key search.* The increases in computer power have made single DES encryption susceptible to hackers. Recently, a group on the Internet issued a $10,000 challenge to break DES.

Triple DES

Triple DES consists of replacing each single encryption with an encrypt, a decrypt, and then a final encrypt. This effectively increases DES security by a multiple of 3. Because the key lengths are doubled and multiplied by 3, an exhaustive key search is much more difficult and security is increased.

RSA

The RSA algorithm was invented in 1978, shortly after DES, by Ron Rivest (R), Adi Shamir (S), and Leonard Adleman (A). The algorithm behind RSA public key encryption is strong and in wide commercial use. RSA is based on public and private key usage.

The U.S. government regulates the import/export of key codes. While it is not illegal to export RSA's mathematical description out of the United States, it is illegal to export an implementation of this cipher.

RPK

RPK is a new encryption method that uses public and private keys developed in New Zealand in the early 1990s. The creators of RPK believe it is an improvement over existing types of public key cryptosystems.

RPK uses an underlying technology that incorporates established cryptographic axioms with new ideas to produce a fast, secure, flexible system. At this time, RPK is fairly new and untested in the smart card field.

Elliptic Curve Cryptosystems (ECC)

Elliptic curve cryptosystems (ECC) are public key cryptosystems, such as RSA and RPK, in which modular multiplication is replaced by the elliptic curve addition operation. An elliptic curve operation involves a sequence of elliptic curve additions, and each addition consists of several arithmetic operations in the finite field. Other public/private key systems utilize exponentiation involving a sequence of modular multiplications rather than the elliptic curve addition operations.

The curves used in elliptic curve analogs are discrete logarithms. These discrete computations are normally of the following form:

$$y^2 = x^3 + ax + b \pmod{p}$$

where p is prime

The problem tapped by the discrete logarithm analogs in elliptic curves is defined as follows:

> Given a point G on an elliptic curve with order r (number of points on the curve) and another point Y on the curve, find a unique x (0 x $r-1$) such that $Y = xG$ (i.e., Y is the xth multiple of G).

The methods have an approximate average running time of a constant times the square root of r, which is much slower than specialized attacks on certain types of groups. The lack of specialized attacks means that shorter key sizes for elliptic cryptosystems give the same security as larger keys in cryptosystems that are based on the discrete logarithm problem.

ECC has recently come into strong consideration, particularly by standards developers, as alternatives to established standard cryptosystems such as the RSA cryptosystem and cryptosystems based on the discrete logarithm problem, including Diffie-Hellman and the Digital Signature Standard (DSS). Some are calling elliptic curve cryptosystems the next generation of public key cryptography, providing greater strength, higher speed, and smaller keys than established systems.

SET Protocol

SET relies on public key cryptography to ensure message confidentiality. In SET, message data will initially be encrypted using a randomly generated symmetric encryption key. This key, in turn, will be encrypted using the message recipient's public key. This is referred to as the *digital envelope* of the message and is sent to the recipient along with the encrypted message itself. After receiving the digital envelope, the recipient decrypts it using the private key to obtain the randomly generated symmetric key and then uses the symmetric key to unlock the original message.

SET uses a distinct public/private key pair to create the digital signature. Thus, each SET participant will possess two asymmetric key pairs: (1) a key exchange pair, which is used in the process of encryption and decryption, and (2) a signature pair for the creation and verification of digital signatures. Note that the roles of the public and private keys are reversed in the digital signature process, where the private key is used to encrypt or sign and the public key is used to decrypt or verify the signature.

Other Cryptography Methods

There remain other crypto companies that have developed workable technologies. As technology moves forward, cryptography will continue to evolve and adopt increasingly more advanced methods. For example, a new public key cryptosystem, NTRU is a system based on manipulating very small integers, thus allowing higher speeds while minimizing computing power. This is one of the new systems not based on factorization or discrete logarithm problems.

IBM supplies a product named Cryptolope. It is a key-based system for securing payments, transactions, and information over the Internet.

In summary, the encryption algorithm is the most integral aspect of the security system and must be selected with the utmost care.

THE SMART MANAGER'S
INDUSTRY SOURCEBOOK

Associations

Advanced Card Technology Association of Canada
7 Lies Street, Ajax, Ontario, Canada L1T 3V7
Phone: 905 683 1442   Fax: 905 683 1033

Federal Smart Card Users Group
Phone: 202 874 7611   Fax: 202 874 8861

International Card Manufacturer Association
34-C Washington Road, Princeton Junction, NJ 08550
Phone: 609 799 4900   Fax: 609 799 7032

Java Card Forum
Executives:
Christian Goire, President
Phone: 33 1 39 46 04   E-mail: Christian.Goire@Bull.net
Michel Roux, President Business Committee
Phone: 33 4 42 36 56 54   E-mail: michel.rouxx@gemplus.com
Bertrand du Castel, President Technical Committee
Phone: 1 512 331 3205   E-mail: ducastel@slb.com

SCIA
191 Clarkville Rd, Lawrenceville, NJ 08550
Phone: 800 848 SCIA   Fax: 609 799 7032
Web Address: scia@scia.org

Smart Card Forum
8201 Greensboro Drive Suite 300, McLean, VA 22102
Phone: 703 610 9023   Fax: 703 610 9005

Smart Card Forum (Germany)
Luner Rennbahn 7, Lüneburg, Germany 21339

The Smart Card Club (UK)
8-9 Bridge Street, Cambridge, England CB2 1UA
Phone: 440 122 332 9900   Fax: 440 122 335 8222

Card Issuers

American Express Travel Related Services
American Express Service Centre, 210 North 2100 West, 2nd
    Floor, Salt Lake City, UT 84116
Phone: 801 965 5003   Fax: 801 965 5270

Banksys
Chaussee de Haecht 1442 Bruxelles, Belgium 1130
Phone: 32 27 276 25 5   Fax: 32 27 276 76 7

CyberMark
PO Box 666458, Tallahassee, FL 32313-6458
Phone: 904 561 1055   Fax: 904 561 1120

Danmont
Ringager 2, Brondby, Denmark DK-2605
Phone: 45 434 43333   Fax: 45 434 49030

MasterCard
World Financial Center, 200 Vessey Street, 31st Floor, New York,
   NY 10285-3100
Phone: 212 640 3095   Fax: 212 619 9765

Mondex International, Ltd.
47-53 Cannon Street, London, England EC4M 5SQ
Phone: 44 0171 5575000   Fax: 44 0171 5575200

Mondex USA
111 Pine Street, 6th Floor, San Francisco, CA 94111
Phone: 415 396 6639   Fax: 415 975 7085

MTA Card Company
347 Madison Ave., New York, NY 10017
Phone: 212 878 0128   Fax: 212 878 0143

Net 1
4th Floor, North Wing, President Place, Rosebank, South Africa
Phone: 27 11 880 5850   Fax: 27 11 880 7080

Prosys International Limited, Inc.
Antel Corporate Center, Suite 1001, 139 Valero, Lagaspe Village
   Makati City, Manila, Philippines 1226
Phone: 62 2 750 0980   Fax: 63 2 750 0986

SEMP/VISA Espana
Gustavo Fdel Balbuena, 15, Madrid, Spain 28002
Phone: 34 1 34 65 30 0   Fax: 34 1 34 65 44 3

Sentec Oy
PO Box 31, Vantaa, Finland FIN 01741
Phone: 358 98 941 1   Fax: 35 898 786 133

Visa International
PO Box 8999, San Francisco, CA 94128-8999
Phone: 415 432 3460   Fax: 415 432 3199

Western Governor's Association
600 17th Street, Suite 1705 South Tower, Denver, CO 80202
Phone: 303 623 9378   Fax: 303 434 7309

Card Manufacturers

Allegheny Printed Plastics
1224 Freedom Rd., Clanberry Township, PA 16066
Phone: 800 933 4123   Fax: 412 776 2909
Web Address: www.allegheny.com
Description: Manufactures print overlaminated cards for credit,
    financial, and access industries.

Allsafe Co.
1105 Broadway, Buffalo, NY 14212
Phone: 716 896 4515   Fax: 716 8964 241
Web Address: www.allsafe.com
Description: Manufactures PVC and polyester cards of various
    technologies, including magnetic stripe, smart card, bar code.

AmaTech GmbH & Co. KG
Rossbergweg 2, Pfronten, Germany D-87459
Phone: 49 8363 91050   Fax: 49 8363 73265
Description: Provides transponders for contactless cards.

American Bank Note Card Company
200 Park Avenue, New York, NY 10166-4999
Phone: 212 557 9100   Fax: 212 338 0728
Description: Secure dependent company that designs, produces,
    and personalizes cards.

American Bank Note Holographics, Inc.
399 Executive Boulevard, Elmsford, NY 10523
Phone: 914 592 2355   Fax: 914 592 3248
Description: Researches, develops, manufactures, and markets
  holograms and laminates for security and authentication
  purposes.

American Magnetics Corporation
740 Watsoncenter Road, Carson, CA 90745-4181
Phone: 213 775 8651   Fax: 310 834 0685
Description: Manufactures terminal equipment.

American Microdrive Manufacturing, Inc.
3140 De La Cruz Blvd., Santa Clara, CA 95054-2046
Phone: 408 986 1122   Fax: 408 986 1121
Web Address: www.ammismarteards.com

Arthur Blank & Co.
225 Rivermoor Street, Boston, MA 02132
Phone: 617 325 9600   Fax: 617 325 1235
Description: Provides plastic products to cover cards.

Aspects Software, Ltd.
Bearford House, 39 Hanover St., Edinburgh, Scotland EH2 2PJ
Phone: 44 131 225 9900   Fax: 44 131 225 9955
Description: Tests and develops equipment for the card terminal
  interface.

BA Custom Cards
5915 Wallace St., Mississauga, Ontario, Canada L4Z I28
Phone: 905 712 4500   Fax: 905 712 4505
Description: Manufactures conventional and smart cards;
  personalization and fulfillment services.

Bull Personal Transaction Systems
300 Concord Road, Billerica, MA 01821
Phone: 508 294 3670   Fax: 508 294 3671
Web Address: www.cp8bull.net
Description: Markets highly scarce microprocessor-based smart
    cards, related terminals.

Cardcorp PTY, Ltd.
40 Brodie St. Rydalmere, Sydney, NSW, Australia
Phone: 61 2 9898 0922   Fax: 61 2 9898 0108
Description: All types of plastic cards; Visa and MasterCard smart
    cards.

Cardlogix
150 Paularino Suite 276, Costa Mesa, CA 92626
Phone: 714 437 0587   Fax: 714 437 0589
Description: Offers a wide range of products from standard smart
    cards to proprietary designs; specifically engineered smart
    cards for the banking, health care, telecommunications,
    entertainment, and gambling industries.

Cardtech, Inc. (a division of Giesecke & Devrient, America)
2020 Enterprise Parkway, Twinsburg OH 44087-1028
Phone: 216 425 1515   Fax: 216 429 9105
Description: Plastic card manufacturer—Visa/MasterCard
    certified; personalized customer services.

Colorado Plasticard, Inc.
10368 W. Centennial Road, Littleton, CO 80127
Phone: 303 973 9311   Fax: 303 973 8420
Description: Visa and MasterCard products along with card
    insurance service bureau capabilities.

Dai Nippon Printing
BF Systems Development Division 3-6-21, Nishi-Gotanda,
    Shinagawa-Ku, Tokyo, Japan
Phone: 81 3 5496 6810   Fax: 81 3 5496 6819
Description: Secure card manufacturer.

DataCard (See Gemplus)

Datakey, Inc.
407 W. Travelers Trail, Burnsville, MN 55337
Phone: 612 890 6850   Fax: 612 890 2726
Web Address: www.datakey.com
Description: International supplier of electronic products and
    services provides product, subsystem, and system solutions for
    recording, storage, and secure transmission of electronic
    information.

De La Rue Card Technology, Ltd.
Ashchurch Business Center-Alexandria Way, Tewkesbury, UK
    GL20 8NB
Phone: 44 1684 290 290   Fax: 44 1684 290 111
Description: Supplies a full range of IC cards and magnetic stripe
    cards.

De La Rue Faraday
4250 Pleasant Valley Rd., Chantilly, VA 20110-5064
Phone: 703 263 0100   Fax: 703 263 2008
Description: Complete plastic card service bureau. Embossing,
    encoding, thermal printing, customer selected PINs, matching
    to carriers and mailing. Visa and MasterCard approved.
    Virginia, Illinois, and California facilities.

Deister Electronics
9303 Grant Ave., Manassas, VA 20110-5064
Phone: 703 368 2739   Fax: 703 368 9791
Web Address: www.deister@deister.com
Description: Manufactures high-security, field-programmable
    proximity cards and readers.

Delphic Card Systems EEIG
Ashchurch Business Centre-Alexandria Way, Tewkesbury, UK
    GL20 8NB
Phone: 44 1684 290 290   Fax: 44 1684 290 1111
Description: Supplies a full range of ID cards, products, services,
    and technical consultancy.

Deluxe Card Services
1050 W. County Road F, Shoreview, MN 55126
Phone: 800 842 4712   Fax: 612 481 4043
Description: Plastic card solutions including ATM, credit, debit, and photo cards; Visa/MasterCard certified.

Didier
PO Box 10748, 613 High Street, Fort Wayne, IN 46853
Phone: 800 852 7373   Fax: 219 424 4923
Description: Manufactures secure cards.

Dittler Brothers, Inc.
1375 Seaboard Ind. Blvd., Atlanta, GA 30318
Phone: 404 355 3423   Fax: 404 355 2569
Description: Specialty printer of promotional devices, including phone cards and game cards.

DocuSystem Corporation
8255 N. North Central Park, Skokie, IL 60079-2970
Phone: 874 329 6584   Fax: 615 254 9471
Description: Card and ticket manufacturer.

Fabrica Nacional de Moneda & Timbre (FNMT)
C/Jorge Juani, 106, Madrid, Spain 28009
Phone: 341 566 6666   Fax: 341 566 6559
Web Address: www.fnmt.es
Description: Smart cards, electronic purses, and development of operating systems.

Gemplus
B P 100-1388 1, Gemenos Cedex, France
Phone: 33 4 42 36 50 00   Fax: 33 4 32 36 50 90
Web Address: www.gemplus.com
Description: Smart and plastic card manufacturing, personalization, and solutions.

(North America) 101 Park Drive, Montgomeryville PA 18936
Phone: (North America) 215 654 8535   Fax: 215 654 9450

Giesecke & Devrient America, Inc.
11419 Sunset Hills Road, Reston, VA 20190-5207
Phone: 703 709 5828   Fax: 703 471 5312
Description: Provides smart card applications and systems; card
    manufacturer.

GPT Card Technology
Card Technology House, New Century Park, Box 53, Coventry
    England CV3 1HJ
Phone: 44 120 356 4550   Fax: 44 120 356 4540
Web Address: www.gpt.co.uk
Description: Division of GPT Payphone Systems, makes smart
    phone cards.

Identicard Systems
40 Citation Lane, PO Box 5349, Lancaster, PA 17606
Phone: 717 569 5797   Fax: 717 569 2390
Description: Manufacturer of security ID cards, access control,
    video-imaging systems.

Innovative Plastic Printing Corp.
445 Gundersen Drive, Carol Stream, IL 60188
Phone: 630 665 0003   Fax: 630 665 7752
Description: Plastic cards, including magnetic stripes and smart
    cards.

Interlock AG
Ruetistasse16, Schlieren, Switzerland, CH-8952
Phone: 44 1 730 80 55   Fax: 41 1 730 8046
Description: Smart cards/contactless cards/smart cards combined
    to any other coding/image software.

International Plastic Cards, Inc.
366 Coral Circle, El Segundo, CA 90245
Phone: 310 322 4472   Fax: 310 322 3489
Description: Plastic card services, including design, manufacturing, processing with quick turnaround, and security.

Kirk Plastics (See Oberthur Smart Cards USA)

Laminex (D&K Laminex, Inc.)
8350 Arrowridge Blvd., Charlotte, NC 28273
Phone: 704 679 4170   Fax: 704 679 8490
Web Address: www.laminex.com
Description: Manufacturer of photo identification systems, cards, and badges.

Landis & Gyr Communications
70 Grand Pre, Geneva 2, Switzerland CH 1211
Phone: 41 22 749 3355   Fax: 41 22 733 5219
Description: Chip cards, optical cards, magnetic cards, card-based payment systems, pay phones, pay phone systems, access control, security modules.

LaserCard Systems Corp.
2644 Bayshore Parkway, Mountain View, CA 94043
Phone: 415 969 4428   Fax: 415 967 6524
Web Address: www.lasercard.com
Description: LaserCard optical memory card, optical card drives, device drivers, and custom software/technical program.

Melzer Mashinenbau GmbH
Ruhrstrasse 51-55, Schwelm, Germany D-58332
Phone: 49 2336 9292 80   Fax: 49 23 36 92 92 85
Description: Manufacturer of machines for producing smart and contactless cards.

Muhlbauer High Tech International
Werner von Siemens Str. 3, Roding, Germany D-93426
Phone: 49 9461 952-0   Fax: 49 9461 952-101
Description: Product automation for semiconductors and smart
  card industry.

National City Processing Company
1231 Durrett Lane, Louisville, KY 40258
Phone: 502 423 3800   Fax: 502 423 3808
Description: Transaction processor.

Nagase California Corp.
Phone: 408 773 0700   Fax: 408 773 9567
Description: Card bodies, smart card module assembly service,
  Toshiba modules, cards, reader/writer.

NBS Technologies Inc.
171 Webster Road, Kitchener, Ontario, Canada N2C 2E7
Phone: 519 893 4510   Fax: 519 748 9843
Web Address: www.nbstech.com
Description: Manufacturer of credit card imprinters.

NEC Card Services
2307 Directors Row, Indianapolis, IN 46241
Phone: 317 227 7700   Fax: 317 227 7799
Description: Plastic card issuance and personalization service.

Nexcom Technology, Inc.
532 Mercury Drive, CA 94056
Phone: 408 730 3690   Fax: 408 720 92~8
Description: Serial flash memory cards; removable flash memory
  solutions.

Norprint USA
1827 Power Ferry Road, Atlanta, GA 30339
Phone: 770 218 0321   Fax: 770 218 0318
Description: Debit cards, transit cards and tickets, ID cards, phone
    cards.

Oberthur Smart Cards USA (Kirk Division)
281 Ana St., Rancho Dominguez, CA 90221
Phone: 310 884 7900   Fax: 310 884 7901
Web Address: www.kirkplastic.com
Description: Memory chip, smart cards, card personalization.

ODS R. Oldenbourg Datansysteme GmbH & Co. KG
Ludwing-Erchard 16 Neufahrn, Germany D-8 5375
Phone: 49 8165 930 0   Fax: 49 8165 930 202
Description: Production of chip cards, full service (consulting,
    design, personalization, lettershop), software, and hardware
    related to smart cards.

OF Orell Fussli, Security Documents
Dietzingerstrasse 3 Postfach, Zurich Switzerland CH-8038
Phone: 41 1 46677 11   Fax: 41 1 4667286
Description: Personalization of credit and debit and member cards
    on magnetic strips and smart cards.

Ordacard
42 Hagilgal Street, Ramat-Gan, Israel 52392
Phone: 972 537 99021   Fax: 972 357 98461
Description: Card manufacturer.

Orga Card Systems
Station Square Two, 2nd Floor, Paoli, PA 19301
Phone: 610 993 9810   Fax: 610 993 8641
Web Address: www.orga.co.uk
Description: Microprocessor-based smart cards for wireless
    phones, banking/retail, and health care applications.

Orga Kartensysteme GmbH
An der Kapelle 2, Paderborn, Germany 33104
Phone: 49 52 54 9 91-0   Fax: 49 52 54 9 91-2 99
Web Address: www.orga.com
Description: Consultant and manufacturer of microprocessor-
    based cards.

Perfect Plastic Printing Corp.
345 Kautz Road, St. Charles, IL 60510
Phone: 630 584 1600   Fax: 630 584 0648
Description: Plastic card printing.

Personal Cipher Card Corp.
3211 Bonnybrook Drive N., Lakeland, FL 33811
Phone: 941 644 5026   Fax: 941 644 1933
Description: Developer of a smart card operating system; owner of
    a variety of masks with leading IC manufacturers supporting
    memory configurations, licenses to operating system with smart
    card manufacturers.

Philips Smart Cards & Systems Inc.
16 New England Executive Park, Burlington, MA 01803
Phone: 617 238 3463   Fax: 617 238 3466
Description: Microprocessor cards, memory cards, readers,
    software.

Plastag Corporation
5990 Northwest Highway, Chicago, IL 60631
Phone: 314 792 2890   Fax: 312 792 3544
Description: Value card, identification, and smart card
    manufacturer.

Plastic Graphics
1710 Cordova St., Los Angeles, CA 90007
Phone: 213 737 0397; 800 553 4611   Fax: 213 737 7236
Description: Manufacturer of plastic cards.

Racom Systems, Inc.
6080 Greenwood Plaza Boulevard, Englewood, CO 80111
Phone: 303 771 2077   Fax: 303 771 4708
Description: Contactless card supplier.

Schlumberger Cards & Systems
50 Ave. Jean Jaures, BP 620-12, Montrouge, France 92542
Phone: 33 1 47 46 66 67   Fax: 33 1 47 46 63 67
Web Address: www.slb.com/et
Description: Smart cards, magnetic stripe cards.

Schlumberger Malco, Inc.
9800 Reisterstown Rd., Owings Mills, MD 21117
Phone: 410 363 1600   Fax: 410 363 4336
Web Address: www.slb.com/et
Description: Secure magnetic stripe cards, smart cards, magnetic
   tape.

Schlumberger Smart Cards & Systems-North America
1601 Schlumberger Drive, Moorestown, NJ 08057
Phone: 609 234 8000   Fax: 609 234 7178
Web Address: www.slb.com/et
Description: Smart cards, secure magnetic stripe cards, bank and
   retail terminals, self-service POS terminals, parking and transit
   products.

Security Card Systems, Inc.
399 Denison Street, Markhan, Ontario, Canada L3R 1B7
Phone: 905 475 1333   Fax: 905 475 5107
Description: Produces magnetic stripe and chip card products.

Sempac SA
Hinterberg Strasse 9, Cham, Switzerland 6330
Phone: 41 41 749 5353   Fax: 41 41 741 6124
Description: Chip card assembly lines.

Sillcocks Plastics International, Inc.
310 Snyder Ave., Berkeley Heights, NJ 07922
Phone: 908 665 0300   Fax: 908 665 9254
Description: Manufactures contact and contactless smart cards.

Spartanics
3605 Edison Place, Rolling Meadows, IL 60008-1077
Phone: 847 394 5700   Fax: 847 394 0409
Web Address: www.spartanics.com
Description: Card-blanking and card-counting equipment.

SSI Custom Data Cards
1027 Waterwood Parkway, Edmond, OK 73034
Phone: 405 359 6000   Fax: 405 359 6528
Web Address: www.ssiphoto.com
Description: Bar code, magnetic stripe, and laminated polyester in
   a wide variety of configurations.

SuperTech Systems Inc.
2425 North Central Expressway, Suite 400, Richardson, TX 75080
Phone: 972 231 2037   Fax: 972 231 2041
Web Address: www.supertecsystems.com
Description: Smart card systems integration, card manufacturing,
   reader/writer, development and manufacturing, smart card
   application system development.

Toshiba America Information System, Inc.
9740 Irvine Boulevard, PO Box 19724, Irvine, CA 92713-9724
Phone: 714 583 3829   Fax: 714 583 3253
Description: Card and terminal manufacturer.

US 3 Inc.
1615 Wyatt Drive, Santa Clara, CA 95054
Phone: 408 748 7725   Fax: 408 748 7724
Description: Largest U.S.-based plastic card producer.

Veron SpA
Via Caldera 21 Milan, Italy 20153
Phone: 39 2 48 21 51   Fax: 39 2 48215 226
Description: Smart card with contact and contactless/desktop and
    portable payment terminals (EFT-POS).

**Versatile Card Technology, Inc. (formerly University Printing
    Services, Inc.)**
5200 Thatcher Road, Downers Grove, IL 60515
Phone: 630 852 5600   Fax: 630 852 5817
Web Address: www.versacard.com
Description: Manufacturer of plastic cards.

Worldtronix
2039 rue du Pont, Marieville, Quebec, Canada J3M 1J8
Phone: 514 460 7411   Fax: 514 460 7073
Web Address: www.worldtronix.ca
Description: Contact-contactless smart card manufacturing.

Semiconductor Manufacturers

Hitachi
200 Sierra Point Parkway, MS080, Brisbane, CA 94005-1835
Phone: 415 589 8300   Fax: 415 583 4207
Web Address: www.hitachi.com

Motorola
432 North 44th Street, Phoenix, AZ 85008
Phone: 602 302 8129   Fax: 602 302 8055

NEC
1201 New York Avenue, NW, Suite 1200, Washington, D.C. 20005
Phone: 202 408 4762   Fax: 202 408 4791
Web Address: www.netech.com

Philips Semiconductors
Mikron GmbH, Gratkorn, Austria A 8101
Phone: 43 4 331 24299950   Fax: 43 433 124299270

Siemens
PO Box 801709, Munich, Germany D 81617
Phone: 43 894 144 25 57   Fax: 49 894 144 22 14

Siemens
10950 North Tantau Ave., Cupertino, CA 95014
Phone: 408 895 5147   Fax: 408 895 5180

Siemens Automation
Würzburger Strasse 121, Entrance Breslauer Strasse, Fürth,
    Germany D 90766
Phone: 49 911 750 22 38   Fax: 49 911 750 4737

Siemens Nixdorf Information System, Inc.
5500 Broken Sound Boulevard, PO Box 310706, Boca Raton, FL
    33487
Phone: 561 997 3486   Fax: 561 997 3416

SGS Thomson Microelectronics
Z.I. de Rousset-B.P. 2, Rousset Cedex, France 13106
Phone: 33 3 34 22 58 800   Fax: 33 3 34 22 58 729

Texas Instruments
12501 Research Boulevard, MS 2201, Austin, TX 78714
Phone: 512 250 4021   Fax: 512 250 7104

Toshiba
9740 Irvine Boulevard, PO Box 19724, Irvine, CA 92713-9724
Phone: 714 583 3829   Fax: 714 583 3253

System Integrators

Applied Systems Institute, Inc. (de La Rue)
1420 K Street, NW Suite 400, Washington, D.C. 20005
Phone: 202 371 1600   Fax: 202 789 0335

Dreifus Associates, Ltd. (DAL)
PO Box 915746, 801 West State Road 436 Suite 2035, Longwood,
   FL 32791
Phone: 407 865 5477   Fax: 407 865 5478

Open Domain Integrated Solutions
39175 Liberty Street Suite 229, Fremont, CA 94538
Phone: 408 492 2037 ext. 22037   Fax: 408 492 3125

Phoenix Planning & Evaluation, Ltd.
3204 Tower Oaks Blvd., Rockville, MD 20852
Phone: 301 984 4210   Fax: 301 984 7510

Terminal Manufacturers

ActionTec Electronics, Inc.
1269 Innsbruck Drive, Sunnyvale, CA 94089
Phone: 408 752 7700   Fax: 408 451 9003

American Magnetics Corporation
740 Watsoncenter Road, Carson, CA 90745
Phone: 213 775 8651   Fax: 310 834 0685

ASCOM MONTEL
Rue Clade Chappe BP 348, Cedex, France 07503
Phone: 33 3 37 58 14 141   Fax: 33 3 37 58 14 300

Bull Worldwide Information Systems
1211 Avenue of the Americas, New York, NY 10036
Phone: 212 719 0671   Fax: 212 719 0513

CP8 Transac
68 route de Versailles BP 45, Louveciennes Cedex, France 78431
Phone: 33 1 39 66 4516   Fax: 33 1 39 66 4402

CUBIC
1600 Spring Hill Road, Suite #100, Vienna, VA 22182
Phone: 703 883 8990   Fax: 703 883 9678

Dassault AT
110 East 59th Street, New York, NY 10022
Phone: 212 909 0550   Fax: 212 909 0555

Datacard Corporation
5929 Baker Road, #400, Minnetonka, MN 55345
Phone: 612 933 1223

Elcotel, Inc.
6428 Parkland Drive, Sarasota, FL 34243
Phone: 941 758 0389   Fax: 941 755 8595
Web Address: www.elcotel.com

GPT
Edge Lane, Liverpool, UK L79NW
Phone: 4401 120 356 4565   Fax: 4401 120 356 4540

Hypercom, Inc.
2851 West Kathleen Road, Phoenix, AZ 85023
Phone: 602 866 5399   Fax: 602 866 5380
Web Address: www.hypercom.com

IER, Inc.
Suite 140, 4004 Beltline Rd., Dallas, TX 75244
Phone: 214 991 1895   Fax: 214 991 1044

Intellect Holdings Limited
1 Brodie Hall Drive, Bentley, Australia WA 6102
Phone: 408 467 0374   Fax: 61 441263741339

International Verifact-IVI
1621 Cedar Circle Drive, Crestwood, KY 40014
Phone: 502 241 2717   Fax: 502 241 2718

Keycorp
Levil 9, 67 Albert Avenue, Chatswook, Australia NSW 2067
Phone: 612 415 2900   Fax: 612 415 1363

Landis & Gyr
rue de Grand-Pre 70, Geneva 2, Switzerland CH-1211
Phone: 41 22 7493378   Fax: 41 22 7493500

Nortel
8200 Dixie Rd., Brampton, Ontario, Canada L6T4B8
Phone: 905 452 2727   Fax: 905 452 2187

OKI Advanced Products
500 Nickerson Road, Marlborough, MA 01752
Phone: 508 460 8632   Fax: 508 460 8617

Protel Inc.
4150 Kidron Road, Lakeland, FL 33811
Phone: 800 925 8881   Fax: 813 646 5855

Scientific Atlanta, Inc.
4261 Communication Drive, Atlanta, GA 30093-2990
Phone: 770 903 2990   Fax: 770 903 6292

Siemens Automation
Würzburger Strasse 121, Entrance Breslauer Strasse, Fürth,
    Germany D 90766
Phone: 49 911 750 22 38   Fax: 49 911 750 47 37

Toshiba America Information Systems, Inc.
9740 Irvine Boulevard, PO Box 19724, Irvine, CA 92713-9724
Phone: 714 583 3829   Fax: 714 583 3253

VeriFone
Three Lagoon Drive, Redwood City, CA 94065-1561
Phone: 415 598 5694   Fax: 415 598 5504

Internet Information Sites

SMARTCARD PAGE
   ct177.nectec.or.th/~nopporn/smartcard/smartcard.html
Describes the anatomy of smart card, standard smart card, and picture of chip card, as well as how chip works.

PC/SC WORKGROUP, www.smartcardsys.com/
Workgroup is defining the interface between the PC and a smart card reader attached either internally or externally to PC.

MICROSOFT SECURITY ADVISOR
   www.microsoft.com/smartcard/
Microsoft recently announced the release of a smart card Internet development toolkit. This is the place to get it.

OPENCARD FRAMEWORK, www.nc.com/opencard/
Parties are working on designing a network computing standard to be compatible with smart card technology used for travelers to access the Internet or their Intranet.

SUN MICROSYSTEMS—JAVA
   java.sun.com/products/commerce/
Sun is the creator of Java, which is an interpreter language that can dynamically load applications.

TOUCH TECHNOLOGY INTERNATIONAL
   www.touchtechnology.com/
Company interested in expanding the scope of an electronic purse.

AMERICAN BANKERS ASSOCIATION, www.aba.com/
A group comprised of most major banks in the United States.

EUROPAY, www.europay.com/index.htm.
European credit card association.

PCMCIA, www.pc-card.com/
PCMCIA (Personal Computer Memory Card International Association) is an international standards body and trade association with over 300 member companies that was founded in 1989 to establish standards for integrated circuit cards and to promote interchangeability among mobile computers where ruggedness, low power, and small size were critical.

DATA COLLECTION & IDENTIFICATION ONLINE (AIM)
www.industry.net/c/ orgindex/aim
An excellent source for purchasing data on-line.

THE SMART CARD RESOURCE CENTER (AMERKORE INTERNATIONAL), www.smart-card.com/
A website designed to provide information about smart cards.

SMART CARD FORUM, www.smartcrd.com/
The industry body for the smart card industry. DAL was a founding member.
Smart Card Forum's Consumer Privacy Position
Smart Card Forum 5th Annual Meeting
The Security of Smart Cards
Smart Cards: Seizing Strategic Business Opportunities
A Letter from Jean McKenna, President, Smart Card Forum
What's a Smart Card?
Smart Card Factoids

NEWSPAGE—SMART CARDS
pnp.individual.com/cgi-bin/pnp.BuildIssue
News site that provides information on smart cards.

Smart Card Links

CP8 Transac: www.cp8.bull.net/
3-G International, Inc.: www.3gi.com/main.htm
Aladdin, Smartcard development systems:
   www.aks.com/ase/ase.htm
American Express TRS: www.americanexpress.com/
Andersen Consulting: www.ac.com/
Applied Communications Inc.: www.tsainc.com/
AT&T: www.att.com/
Bank of America: www.bofa.com/
Bellcore: www.bellcore.com/
CardTech/SecurTech: www.ctst.com/
Certicom: www.certicom.ca/
Chase Manhattan Bank: www.chase.com/noframes/home.html
Citibank: www.citibank.com/
DataCard Corporation: www.datacard.com/
Diebold/Interbold: www.diebold.com/
Dove Associates Inc.: www.doveassoc.com/
EDS: www.edsr.eds.com/
First Union National Bank: www.firstunion.com/
Forrester Research: www.forrester.com/
Fujitsu-ICL Systems Inc.: www.fujitsu.com/
Gemplus Card International: www.gemplus.com/
Gilbarco, Inc.: www.gilbarco.com/
Health Card Technologies, Inc.: www.hct.com/
Hewlett-Packard Company: www.hp.com/
Huntington Bancshares: www.huntington.com/
Informix Software: www.informix.com/infmx-cgi/Webdriver
Intellect Electronics, Inc.: www.intellect.com.au/
Los Alamos National Laboratory: www.lanl.gov/external/
MasterCard International: www.mastercard.com/
MCI Telecommunications, Inc.: www.mci.com/
Mellon Bank: www.mellon.com/
Mondex USA: www.mondexusa.com/
Motorola: www.mot.com/

National Association of Federal Credit Unions:
    www.nafcunet.org/
NationsBank: www.nationsbank.com/
NBS Card Services: www.nbstech.com/
OKI Advanced Products: www.oap.oki.com/
ORGA Card Systems: www.orga.co.uk/
Perot Systems Corporation: www.ps.net/
PSI: www.psi-nfo.com/
SGS-Thomson Microelectronics: www.st.com/
Sandia National Laboratories: www.sandia.gov/
The Santa Fe Group: www.santafegroup.com/
Schlumberger Electronic Transactions: www.slb.com/et/
Security Dynamics: www.securid.com/
Shell Oil Products Company: www.shellus.com/
Siemens: www.scn.de/
SPYRUS: www.spyrus.com/
State of Ohio: www.ohio.gov/ohio/
Systems & Computer Technology: www.sctcorp.com/
U.S. Department of Defense: www.dtic.mil:80/c3i/marcard.html
U.S. Postal Service: www.usps.gov/
UBIQ Incorporated: www.ubiqinc.com/
Unisys Corporation: www.unisys.com/
US Bancorp: www1.usbank.com/
US WEST: www.uswest.com/
VeriFone: www.verifone.com/
VISA International: www.visa.com/cgi-bin/vee/main.html?2+0
Wells Fargo Bank: wellsfargo.com/

Additional References

Documents

Microsoft Crypto API FAQ
    www.microsoft.com/workshop/prog/security/capi/cryptfaq-f.htm

Microsoft Zero Administration Initiative for Windows
      www.microsoft.com/windows/platform/info/zawmb.htm
NetPC: www.microsoft.com/windows/netpc/default.htm
PC98 Design Guide: www.microsoft.com/hwdev/pc98.htm

PC/SC Workgroup Members

Bull Personal Transaction Systems: www.bull.com
Gemplus: www.gemplus.com
Hewlett-Packard: www.hp.com
IBM: www.chipcard.ibm.com
Microsoft: www.microsoft.com
Schlumberger: www.slb.com
Siemens Nixdorf: www.sni.de
Sun Microsystems: www.sun.com
Toshiba: www.toshiba.com
Verifone: www.verifone.com

THE SMART ORGANIZATION

Smart Management Team

Assumptions: *Minimum of 25 to 30 million smart cards per year. Product mixture of microcontroller and memory contact/contactless cards in equal proportions.*

Minimum staff level: 16+ general managers (if not already in place).

Organization: Hire from within (if possible) to promote technology and knowledge transfer.

1. **Smart Card Manufacturing Team**
   Smart card manufacturing manager
   Production development supervisor
   Process and quality control supervisor

2. **Personalization Team**
   Personalization manager
   Network/communication analyst

Application programmer/analyst
Process and quality control supervisor

3. **Product Development Team**
   Product development manager
   Card systems analyst (central resource)
   Card integration analyst (8 to 12 projects/year)
   Mask/OS programmer/analyst (4 projects/year)
   Smart card application programmer/analyst (5 projects/year)
   Terminal application programmer/analyst (5 projects/year)
   Integration specialist (central resource)

4. **Project Management/Implementation Team**
   Integration project manager
   Application support manager—vertical market segment 1
   Application support manager—vertical market segment 2
   Application support manager—vertical market segment 3
   Ongoing support manager

5. **Security**
   Security officer (central resource)
   (For example the Visa/MasterCard secure card manager)

Smart Job Descriptions

Organizational tasks are roughly described in the following summary descriptions. These are only high-level guideline skillsets and responsibilities.

This should not be interpreted as the precise team or configuration for a business unit. Some of the tasks are modular (e.g., card manufacturing or software/application development). Each organization shall seek the appropriate size and scale of expertise (e.g., R&D-heavy for complex, new applications or application-support-management-heavy for solutions organizations). The skills can and will often overlap to other team members during the staffing and enabling process for your business.

Smart Manager

Overall enterprise project leader facilitating the development and implementation of smart card business initiative(s) for organization.

General management skills, with discipline understanding of (minimum):

1. Hardware (microcomputer)
2. Software (system and application)
3. Project and process management
4. Distributed computing
5. Security considerations (physical and application specific)
6. Complex project management

Strong communications skills.

Experience 4 to 6 years minimum in information technology business management. Ability to evaluate and manage team of minimum of 15 to 20 diverse disciplines and build cohesive team.

Preferred degrees: Management (MBA and/or information technology discipline study)

1. Smart Card Manufacturing

Smart Card Manufacturing Manager

Production process manager reporting to smart manager to implement production line, oversee production process and procedures, process control, and training. Responsible for managing maintenance team (may or may not be outsourced depending on variety of equipment and company policies).

Experience in operations and complex CNC and computerized factory automation systems. Responsibilities include:

1. Machine utilization and production scheduling
2. Quality control and resource management
3. Training and staff development

4. Growth forecasting and manufacturing planning
5. Profitability and asset control

Production Development Manager

Production development manager reporting to smart manager responsible for implementing and supporting new product development and manufacturing processes and procedures. Responsibilities include:

1. Product definition and requirements specification
2. Production machine qualification and specification
3. Production specifications and qualification of QA/QC procedures
4. Implement statistical testing and other quality control procedures

Process and Quality Control Manager

Reporting to smart manager and production manager, the quality manager is responsible for implementing and continuously improving yield, productivity, and efficiency of smart card manufacturing line(s).

1. Design, implement, and review statistical testing and other quality control enterprise procedures.
2. Enable ISO 9000+ qualification and continuous improvement programs.
3. Enable and oversee third-party review and external expertise.
4. Report and communicate and enable improvements and areas of concern to team and management.

2. Personalization

Personalization Manager

Personalization process manager reporting to smart manager to develop and implement production line, process control, and training. Responsible to manage maintenance team.

Experience in operations and complex personalization and computerized factory automation systems. Responsibilities include:

1. Machine utilization and production scheduling
2. Quality control and resource management
3. Training and staff development
4. Growth forecasting and manufacturing planning
5. Profitability and asset control

Network/Communication Analyst

Network communication analyst reporting to personalization manager to implement and control networks for communication between various databases. Responsibilities include:

1. Design and implementation of networks
2. Efficient and reliable communication
3. Support of data-file conversions and threading of personalization information for all projects

Experience in working with a variety of communication protocols and databases required.

Application Programmer/Analyst

Application programmer/analyst reporting to the personalization manager responsible for developing and implementing application programs for the physical and electrical personalization of smart cards. Responsibilities include:

1. Personalization process definition
2. Card test procedures and QC software development for personalization machines
3. Development and implementation of customer application software
4. Control of batch processing
5. Version and process control software management

Process and Quality Control Manager

Reporting to personalization manager, the quality manager is responsible to implement and continuously improve yield and overall effective productivity and efficiency of smart card personalization line(s).

1. Design and review statistical testing and other quality control procedures.
2. Implement process improvements, working closely with quality manager.
3. Monitor production-line operations.
4. Schedule and oversee line training and periodic recurrent training.

3. Product Development

Product Development Manager

Product development manager reporting to the smart manager responsible to implement and support new product development, including smart card and terminal soft- and hardware. Responsible to manage software and hardware product developments internally. Responsibilities include:

1. Analyze new products (solutions) together with the sales and marketing people.
2. Define product technical and cost requirement specifications.
3. Define test and acceptance specifications.
4. Develop, test, and implement products.
5. Manage the internal development projects of and between the various disciplines.
6. Determine "make versus buy" decisions for application software, mask operating systems, and development/implementation tools.

Card Systems Analyst

Card systems analyst reporting to the product development manager responsible to analyze and design the smart card in relation to the chip, OS, type of card, and smart card application structure. Responsibilities include:

1. Define application smart card requirements.
2. Define smart card application structure.
3. Refine/validate customer specifications and potential data structure inconsistencies.
4. Define personalization and testing procedures.

Card Integration Manager

Card integration manager reporting to the product development manager responsible for the integration process between the card operating system, card application software, and the terminal application software. Responsibilities include:

1. Define and specify interface specifications.
2. Develop component level and subsystem testing procedures to support customer certification requirements (if any).
3. Liaison with customer to assure compatibility (hardware and software).

Mask/OS Programmer Analyst

Mask/OS programmer analyst reporting to the product development manager responsible for the implementation and support of OS specifications and OS developments together with the chip suppliers. Responsibilities include:

1. Define OS requirements for customer applications.
2. Define and specify semiconductor requirements.

3. Support selection (make versus buy) of semiconductor and OS.

4. Manage version and application migration control.

Smart Card Application Programmer/Analyst

Smart card application programmer/analyst reporting to the product development manager responsible for the development and support of software development process for the smart card. Responsibilities include:

1. Define the smart card application software requirements.

2. Develop and test the application software.

3. Manage programming team(s).

Terminal Application Programmer/Analyst

Terminal application programmer/analyst reporting to the product development manager responsible for the development and support of software developments for the terminal. Responsibilities include:

1. Define terminal application software development.

2. Develop and test the application software.

3. Support certification and validation process (if necessary).

4. Manage programming team(s).

Integration Specialist

Integration specialist reporting to the product development manager responsible for the integration between the various software programs. Responsibilities include:

1. Define interface requirements for smart card—terminal.

2. Define interface requirements terminal—host.

3. Review overall system implementation and specifications for possible inconsistencies.

4. Work across skill disciplines to pull together complex projects.

4. Project Management

Integrated Project Manager

The integration project manager is reporting to the smart manager to manage the development and implementation of the projects on the customers' sites. Responsible to control overall costs, time, and efficiency. Responsibilities include:

1. Defining project planning
2. Controlling planning
3. Customer interface
4. Budgeting
5. Controlling costs

Application Support Manager (1 minimum per segment/industry)

The application project managers are reporting to the integrated project manager to manage the development and implementation of projects in vertical segments. Responsibilities include:

1. Customer interface
2. Application refinement and specification modification (as or if necessary)
3. Project control planning
4. Project scheduling and critical-path management

5. Security

Security Officer

Security officer reporting to the smart manager to implement secure system solutions. Responsibilities include:

1. Assure compliance with operating rules and/or procedures that may be imposed by customer or associations.

2. Review/monitor/correct procedures and human resources for possible security problems.

3. Provide periodic external reporting as requested/required by customer or associations.

4. Oversee compliance audits and spot checks to assure water-tightness of security infrastructure and procedures.

5. Develop contingency plans for various security-critical scenarios.

INDEX

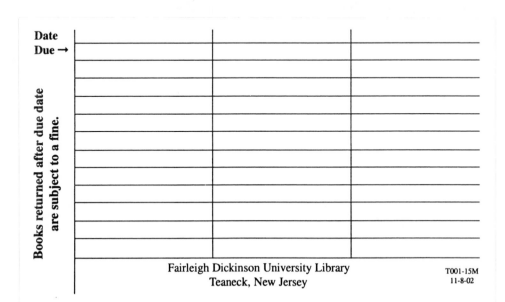